# Holt Algebra 1

## Chapter 12 Resource Book

**HOLT, RINEHART AND WINSTON**

A Harcourt Education Company

**Orlando • Austin • New York • San Diego • London**

Printed in the United States of America

ISBN 0-03-042769-X

1 2 3 4 5 6 7 8 9  862  09 08 07 06

# Contents

Date _______________

Dear Family,

In Chapter 12 your child will learn how to graph rational functions, operate with rational expressions, and solve rational equations.

An **inverse variation** is a relationship that can be written in the form $y = \frac{k}{x}$, where $k \neq 0$. Inverse variation implies that one quantity increases as the other quantity decreases. If you multiply both sides of an inverse variation by $x$ you get $xy = k$, so the product of $x$ and $y$ is constant.

**inverse variation: $y = \frac{12}{x}$**

| $x$ | $y$ | **product** |
|---|---|---|
| 1 | 12 | $1 \cdot 12 = 12$ |
| 3 | 4 | $3 \cdot 4 = 12$ |
| 6 | 2 | $6 \cdot 2 = 12$ |

The graph of an inverse variation has two parts that are not connected. Because the product of $x$ and $y$ is always a nonzero constant, neither $x$ nor $y$ can actually equal zero. That creates lines, called **asymptotes**, that the graph gets close to, but never touches.

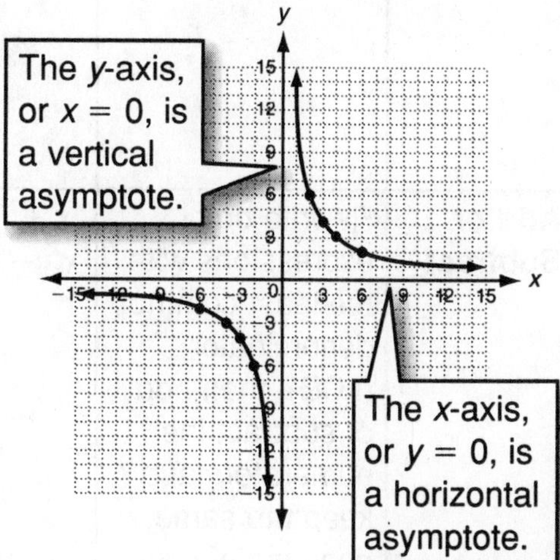


An inverse variation is a special type of **rational function**. A rational function is a function whose rule is a quotient (or ratio) of two polynomials. Because the denominator of a ratio cannot be 0, rational functions have **excluded values** that would make the function undefined.

**Identify the excluded value of $y = \frac{1}{x+2} - 3$.**

$x + 2 = 0$ *Set the denominator equal to 0.*

$x = -2$ *Solve. The excluded value is $-2$.*

Many rational functions can be graphed by translating the parent function $y = \frac{1}{x}$. Most rational functions are discontinuous, meaning their graphs contain one or more jumps, breaks, or holes. Asymptotes are the only discontinuities that your child will learn about in this course.

A **rational expression** is an expression whose numerator and denominator are polynomials. Rational expressions are very much like fractions, and you simplify them in the same way—by dividing out common factors that appear in both the numerator and denominator.

**Simplify $\frac{x^2 + 3x + 2}{x^2 - 1}$.**

$\frac{\cancel{(x+1)}(x+2)}{\cancel{(x+1)}(x-1)}$ *Factor and divide out common factors.*

$\frac{x+2}{x-1}$ *Simplify.*

Simplifying can often be used to divide polynomials by monomials and binomials. Students will learn polynomial long division as well.

The table below shows how to operate with rational expressions.

| Operation | How To | Example |
|---|---|---|
| **Multiply** | Multiply the numerators and multiply the denominators; factor; then divide out common factors. | $\frac{10x-40}{x^2-6x+8}\cdot\frac{x+3}{5x+15}=\frac{(10x-40)(x+3)}{(x^2-6x+8)(5x+15)}$<br>$=\frac{\cancel{(10)}^{2}\cancel{(x-4)}\cancel{(x+3)}}{(x-2)\cancel{(x-4)}\cancel{(5)}\cancel{(x+3)}}=\frac{2}{x-2}$ |
| **Divide** | Multiply by the reciprocal of the divisor. | $\frac{4x^3}{9x^2y}\div\frac{16}{9y^5}=\frac{4x^3}{9x^2y}\cdot\frac{9y^5}{16}$<br>$=\frac{(4)(9)x^3y^5}{(9)(16)x^2y}$<br>$=\frac{\cancel{(4)}\cancel{(9)}\cancel{x^3}^{x}\cancel{y^5}^{y^4}}{\cancel{(9)}\cancel{(16)}^{4}\cancel{x^2}\cancel{y}}=\frac{xy^4}{4}$ |
| **Add or Subtract** | Rewrite both expressions with the least common denominator (LCD); then add or subtract the numerators and keep the same denominator. | $\frac{x+3}{x^2-4}+\frac{5}{x-2}=\underbrace{\frac{x+3}{(x+2)(x-2)}}+\frac{5}{(x-2)}$<br>Identify the least common multiple: $(x+2)(x-2)$<br>$=\frac{x+3}{(x+2)(x-2)}+\underbrace{\frac{5}{(x-2)}\left(\frac{x+2}{x+2}\right)}$<br>Multiply by 1 to get the LCD.<br>$=\frac{(x+3)+(5x+10)}{(x+2)(x-2)}=\frac{6x+13}{x^2-4}$ |

**Rational equations** contain one or more rational expressions. If the equation is a proportion, you can solve it using the Cross Products Property. You can solve *any* rational equation by multiplying both sides of the equation by the LCD of all of the rational expressions. However, you must always look out for **extraneous solutions** that do not check.

**Solve $\frac{3x-7}{x-5}+\frac{x}{2}=\frac{8}{x-5}$.**

$$2\cancel{(x-5)}\left(\frac{3x-7}{\cancel{x-5}}\right)+2\cancel{(x-5)}\left(\frac{x}{\cancel{2}}\right)=2\cancel{(x-5)}\left(\frac{8}{\cancel{x-5}}\right)$$ *Multiply by the LCD and divide out common factors.*

$6x-14+x^2-5x=16$ *Simplify.*

$x^2+x-30=0$ *Set equal to zero.*

$(x-5)(x+6)=0$ *Factor.*

$\cancel{x=5}$ or $x=-6$ *Use the Zero Product Property.*

5 makes two denominators equal to 0, so it is an extraneous solution.

For additional resources, visit go.hrw.com and enter the keyword MA7 Parent.

Name ______________________ Date ____________ Class ____________

LESSON **12-1**

# Practice A
## *Inverse Variation*

**Tell whether each relationship is an inverse relation. Explain.**

**1.**

| $x$ | $y$ | |
|---|---|---|
| 1 | 30 | $xy =$ ________ |
| 2 | 15 | $xy =$ ________ |
| 3 | 10 | $xy =$ ________ |

______________________

**2.**

| $x$ | $y$ | |
|---|---|---|
| 9 | 2 | $xy =$ ________ |
| 7 | 3 | $xy =$ ________ |
| 5 | 4 | $xy =$ ________ |

______________________

**3.** $x + y = 4$

______________________

______________________

**4.** $xy = 7$

______________________

______________________

**5.** Write and graph the inverse variation in which $y = 6$ when $x = 2$ by following the steps below.

**a.** Find the constant of variation.

$k =$ ________

**b.** Write the inverse variation equation.

$y = \frac{\square}{x}$

**c.** Make a table of values and graph.

| $x$ | −6 | −4 | −2 | 0 | 2 | 4 | 6 |
|---|---|---|---|---|---|---|---|
| $y$ | | | | | | | |

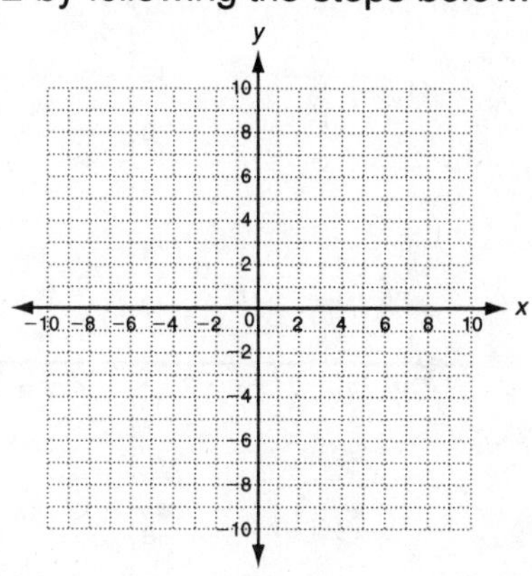


**6.** Let $x_1 = 5$, $y_1 = 4$, and $x_2 = 2$. Let $y$ vary inversely as $x$. Find $y_2$. ________

**7.** Let $x_1 = 3$, $y_1 = 7$, and $y_2 = 21$. Let $y$ vary inversely as $x$. Find $x_2$. ________

**8.** While traveling in a car, the speed of the car $x$ is inversely proportional to the time $y$ it takes to travel a certain distance. At 20 mi/h, it takes 10 minutes to travel to work. How many minutes would it take traveling 25 mi/h?

**a.** List the given information.

$x_1 =$ ________, $y_1 =$ ________, $x_2 =$ ________

**b.** Set up and solve an equation to find $y_2$.

________ = ________

$y_2 =$ ________

It will take ________ minutes.

Name ______________________ Date __________ Class __________

LESSON 12-1

# Practice B

## *Inverse Variation*

**Tell whether each relationship is an inverse relation. Explain.**

1.

| $x$ | $y$ |
|---|---|
| 2 | 12 |
| 3 | 8 |
| 4 | 6 |

________________________

2.

| $x$ | $y$ |
|---|---|
| 1 | 4 |
| 2 | 8 |
| 3 | 12 |

________________________

3. $x = \frac{y}{5}$

________________________

4. $xy = 8$

________________________

5. Write and graph the inverse variation in which $y = 3$ when $x = 2$.

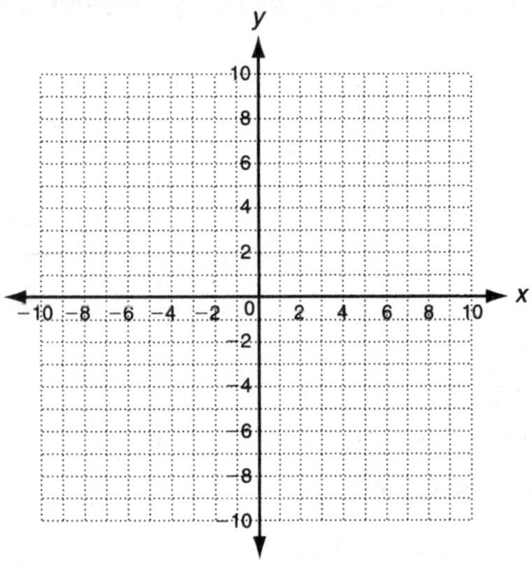

________________________

6. Write and graph the inverse variation in which $y = 1$ when $x = -3$.

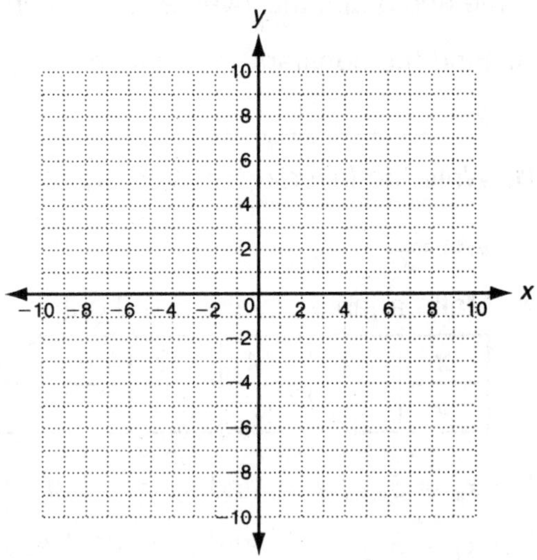

________________________

7. Let $x_1 = 4$, $y_1 = 12$, and $x_2 = 3$. Let $y$ vary inversely as $x$. Find $y_2$. __________

8. Let $x_1 = 3$, $y_1 = 10$, and $y_2 = 15$. Let $y$ vary inversely as $x$. Find $x_2$. __________

9. While traveling in a car, the speed of the car is inversely proportional to the time it takes to travel a certain distance. At 25 mi/h, it takes 15 minutes to travel to work. How many minutes would it take traveling 30 mi/h? __________

10. The amount of pizza that Kirby can buy varies inversely as the price of the pizza increases. Kirby can afford to buy 3 pizzas that cost $15.00 each. How many pizzas that cost $9.00 each can Kirby buy? __________

Name ______________________ Date ____________ Class ____________

LESSON 12-1

# Practice C
## *Inverse Variation*

**Tell whether each relationship is an inverse relation. Explain.**

**1.**

| *x* | *y* |
|---|---|
| 0.25 | 3.5 |
| 0.5 | 1.4 |
| 0.75 | 0.875 |

______________________

**2.**

| *x* | *y* |
|---|---|
| 1 | 3.5 |
| 2.5 | 1.4 |
| 4 | 0.875 |

______________________

**3.** $2x + 4y = 6$

______________________

______________________

**4.** $3xy = 12$

______________________

______________________

**5.** Write and graph the inverse variation in which $y = -4$ when $x = 2$.

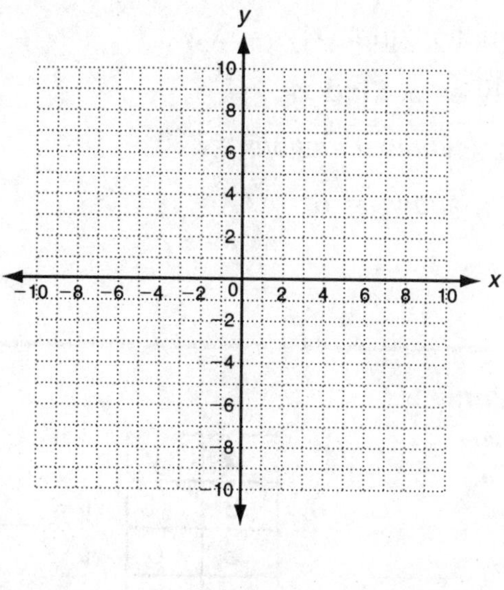


______________________

**6.** Write and graph the inverse variation in which $y = \frac{1}{2}$ when $x = 36$.

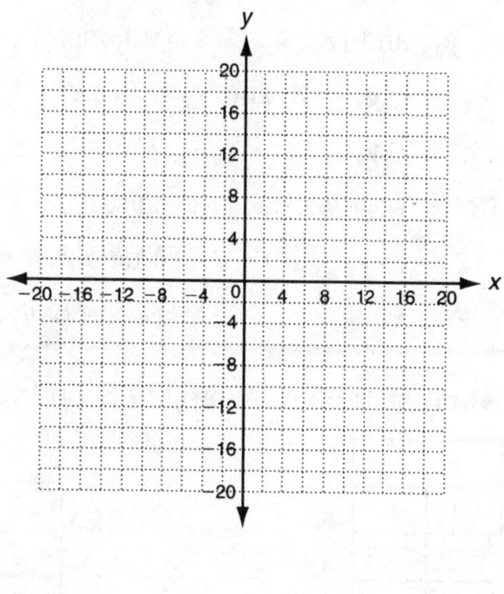


______________________

**7.** Let $x_1 = 8$, $y_1 = 2.5$, and $x_2 = 40$. Let *y* vary inversely as *x*. Find $y_2$. ____________

**8.** Let $x_1 = -1.5$, $y_1 = 4$, and $y_2 = 6$. Let *y* vary inversely as *x*. Find $x_2$. ____________

**9.** The number of hours *y* it takes to paint a house varies inversely with the number of people painting *x*. It takes 2 people 20 hours and 30 minutes to paint a house. How many people would be needed if the house must be painted in 8 hours? ____________

**10.** The volume of a gas *V* varies inversely with the pressure *P*. If a gas with a volume of 30 L and a pressure of 2 atmospheres is compressed to a volume of 25 L, what will be the new pressure? ____________

Name ______________________ Date __________ Class __________

LESSON 12-1

# Reteach

## Inverse Variation

In an **inverse variation**, one quantity increases while the other quantity decreases. There are two ways to tell if a relationship is an inverse variation.

If the product of each ordered pair is constant, then the relationship is an inverse variation.

**Tell whether the relationship in the table is an inverse variation.**

Find $xy$ for each ordered pair.

| $x$ | $y$ | |
|---|---|---|
| 1 | 18 | $1 \times 18 = \mathbf{18}$ |
| 2 | 9 | $2 \times 9 = \mathbf{18}$ |
| 3 | 6 | $3 \times 6 = \mathbf{18}$ |

Because the product is a constant, the relationship is an inverse variation.

If you can write the function rule in $y = \frac{k}{x}$ form, the relationship is an inverse variation.

**Tell whether $xy = 4$ is an inverse variation.**

$xy = 4$

$\frac{xy}{x} = \frac{4}{x}$ *Divide both sides by x.*

$y = \frac{4}{x}$

This function is an inverse relation.

If $(x_1, y_1)$ and $(x_2, y_2)$ are solutions of an inverse variation, then $x_1y_1 = x_2y_2$.

**Let $x_1 = 4$, $y_1 = 3$, and $y_2 = 6$. Let $y$ vary inversely as $x$. Find $x_2$.**

| | |
|---|---|
| $x_1y_1 = x_2y_2$ | *Write the product rule for inverse variation.* |
| $(4)(3) = (x_2)(6)$ | *Substitute $x_1 = 4$, $y_1 = 3$, and $y_2 = 6$.* |
| $12 = 6x_2$ | *Multiply.* |
| $2 = x_2$ | *Divide both sides by 6.* |

**Tell whether each relationship is an inverse variation.**

1.

| $x$ | $y$ | |
|---|---|---|
| 1 | 50 | $xy =$ ______ |
| 2 | 25 | $xy =$ ______ |
| 5 | 10 | $xy =$ ______ |

______________________

2.

| $x$ | $y$ | |
|---|---|---|
| 2 | 12 | $xy =$ ______ |
| 3 | 8 | $xy =$ ______ |
| 4 | 6 | $xy =$ ______ |

______________________

3.

| $x$ | $y$ | |
|---|---|---|
| 3 | 10 | $xy =$ ______ |
| 5 | 6 | $xy =$ ______ |
| 7 | 2 | $xy =$ ______ |

______________________

4. $3x + y = 5$

______________________

5. $4xy = 8$

______________________

6. $x = \frac{y}{10}$

______________________

7. Let $x_1 = 5$, $y_1 = 12$, and $y_2 = 6$. Let $y$ vary inversely as $x$. Find $x_2$. ______________

8. Let $x_1 = 2$, $y_1 = 15$, and $x_2 = 10$. Let $y$ vary inversely as $x$. Find $y_2$. ______________

Name ______________________ Date __________ Class __________

LESSON 12-1

# Reteach

## Inverse Variation (continued)

You can also identify an inverse relation by its graph.

The graph of an inverse relation:
- has two parts that are not connected.
- does not contain (0, 0).

**Write and graph the inverse variation in which $y = 5$ when $x = 2$.**

**Step 1:** Find $k$.

$$k = xy$$
$$= (2)(5)$$
$$= 10$$

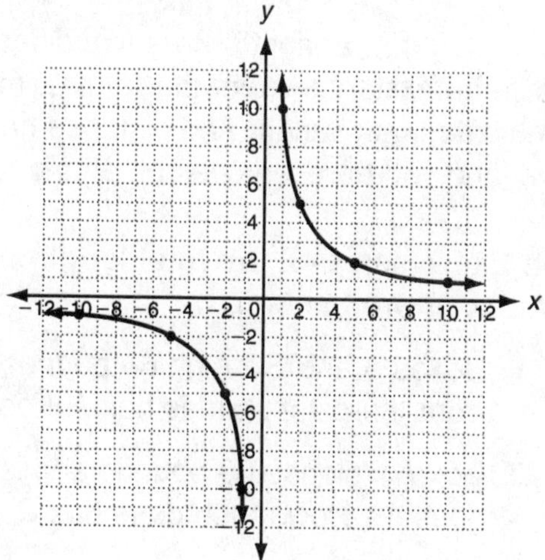

**Step 2:** Write the inverse variation equation.

$$y = \frac{k}{x}$$

$$y = \frac{10}{x}$$

**Step 3:** Make a table of values and graph.

| $x$ | −10 | −5 | −1 | 0 | 1 | 5 | 10 |
|---|---|---|---|---|---|---|---|
| $y$ | −1 | −2 | −10 | und. | 10 | 2 | 1 |

**9.** Write and graph the inverse variation when $y = 3$ and $x = 2$.

**a.** Find $k$. ______________

**b.** Write the inverse variation. ________

**c.** Fill in the table of values and graph.

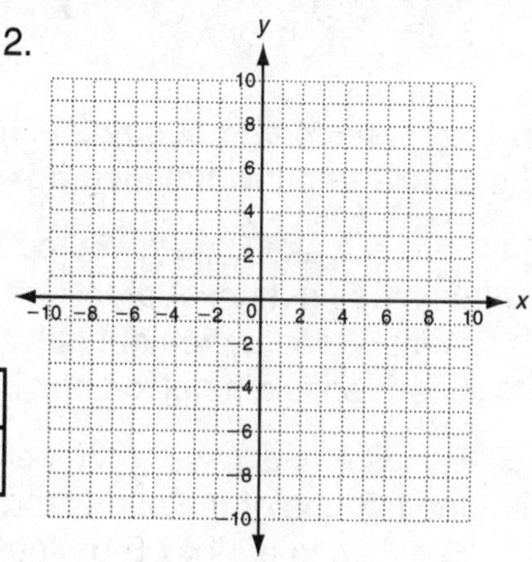

| $x$ | −3 | −2 | −1 | 0 | 1 | 2 | 3 |
|---|---|---|---|---|---|---|---|
| $y$ | | | | | | | |

**10.** Write and graph the inverse variation when $y = 2$ and $x = -4$.

**a.** Find $k$. ______________

**b.** Write the inverse variation. ________

**c.** Fill in the table of values and graph.

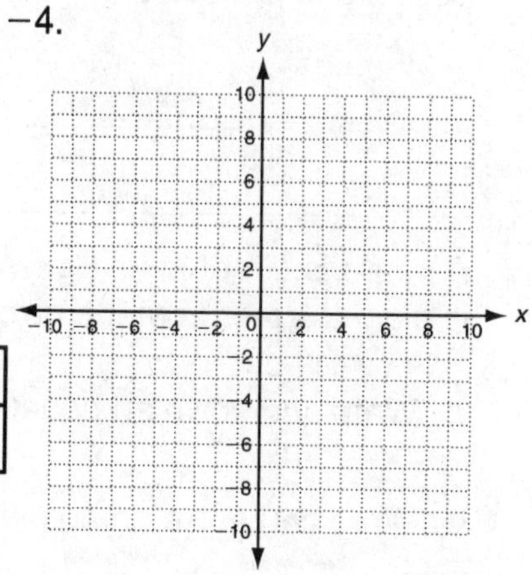

| $x$ | −4 | −2 | −1 | 0 | 1 | 2 | 4 |
|---|---|---|---|---|---|---|---|
| $y$ | | | | | | | |

Name ______________________ Date ____________ Class ____________

LESSON **12-1**

# Challenge
## *Seesaws*

When two children of different weights sit on a seesaw that is joined to the pivot at its midpoint, one child sits higher than the pivot point, and the other child sits lower than it. They achieve balance and reverse who is higher and who is lower by exerting a downward force on the ground.

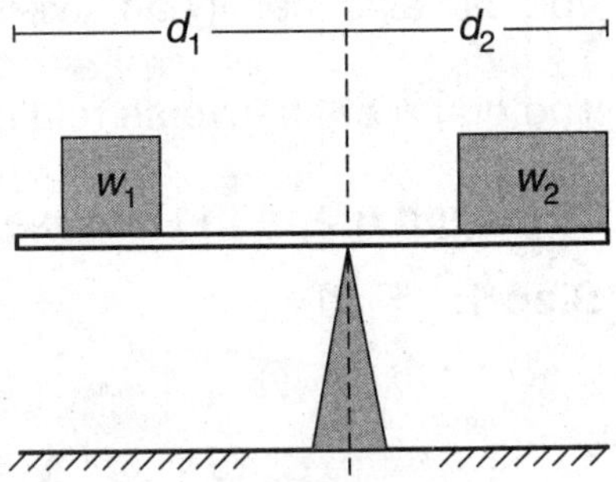


$w_1d_1 = w_2d_2$

The diagram at right shows two different weights positioned at different distances from the pivot point of a seesaw. The equation indicates that weight and distance have an inverse variation. $w_1$ and $w_2$ represent weight, and $d_1$ and $d_2$ represent distance.

**In Exercises 1–3, the total length of a seesaw is 12 feet, and boxes A and B are balanced.**

1. Boxes A and B weigh 60 pounds and 80 pounds, respectively. How far from the pivot point is box A if box B is 4 feet from the pivot point? ____________

2. Box A weighs 100 pounds and is 5 feet from the pivot point of the seesaw. How much does box B weigh if it is 6 feet from the pivot point? ____________

3. Boxes A and B weigh 120 pounds and 150 pounds, respectively. How far is each box from the pivot point of the seesaw? (Hint: Let $x$ represent the distance between box A and the pivot point. Then write and solve an equation for $x$.) ____________

**In Exercises 4 and 5, solve each seesaw problem. The length of the board used to balance boxes A and B is 20 feet.**

4. Boxes A and B weigh 60 pounds and 80 pounds, respectively. John wants to put box A at the left end of the board and to position the pivot point 5 feet to the right of box A. How far from the right end of the board should he place box B? ____________

5. Boxes A and B weigh 60 pounds and 80 pounds, respectively. John wants to put box A at the left end of the board and to position the pivot point 15 feet to the right of box A. Is it possible to put box B on the seesaw so that the boxes balance? Explain your response.

____________________________________________

____________________________________________

6. A board is $L$ feet long. Boxes A and B weigh 60 pounds and 80 pounds, respectively. Box A is placed $x$ feet from the pivot point.

   a. Write and solve an equation for $L$ in terms of $x$. ____________

   b. Write and solve an equation for $x$ in terms of $L$. ____________

Name ______________________ Date __________ Class __________

LESSON 12-1

# Problem Solving

## *Inverse Variation*

**The number of gallons of gasoline that Isaac can afford *y* varies inversely as the price per gallon *x*. Isaac can afford 12 gal when the price is $2.50 per gallon.**

1. Write an inverse variation for this situation.

_______________________

2. Determine a reasonable domain and range.

_______________________

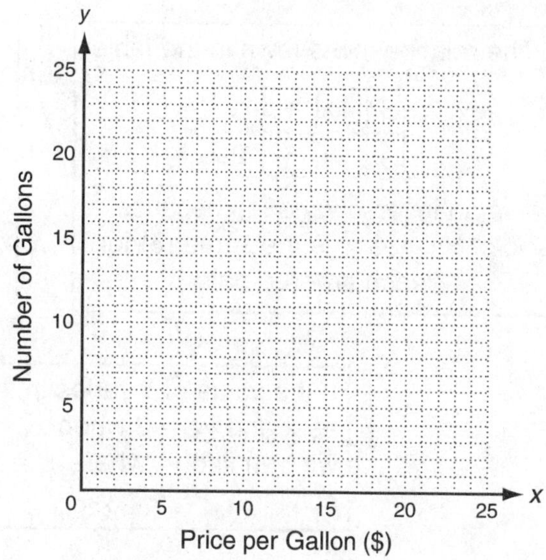


3. Complete this table of ordered pairs and graph the inverse variation.

| *x* | | 2.50 | 3.00 | | 10.00 |
|---|---|---|---|---|---|
| *y* | 20 | 12 | | 6 | |

4. Use the graph to estimate the number of gallons Isaac can afford when the price is $3.49 per gallon.

_______________________

**Select the best answer.**

5. The inverse variation $xy = 9$ relates the current *x* in amps to the resistance *y* in ohms of a circuit attached to a 9-volt battery. What is the resistance of the circuit when the current is 1.8 amps?

**A** 0.2 ohms
**C** 10.8 ohms
**B** 5 ohms
**D** 16.2 ohms

6. Boyle's law states that the pressure of a quantity of gas *x* varies inversely as the volume of the gas *y*. The volume of air inside a bicycle pump is 6.1 in$^3$, and the pressure is 18.3 psi. Which equation represents this situation?

**F** $y = 3x$
**H** $y = 111.63x$
**G** $y = \frac{3}{x}$
**J** $y = \frac{111.63}{x}$

7. The time it takes to drive between two cities varies inversely as speed. Dolores can drive between Houston, Texas, and San Antonio, Texas, in 3.5 h at 60 mi/h. How long will the trip take if she drives 70 mi/h? (Round your answer to the nearest tenth.)

**A** 2.3 h
**C** 3 h
**B** 2.6 h
**D** 4.1 h

8. On a balanced lever, weight varies inversely as the distance from the fulcrum. Angie weighs 120 lb. When she sits $2\frac{1}{2}$ ft from the fulcrum of a seesaw, she balances with her brother, who weighs 80 lb. How far is her brother from the fulcrum of the seesaw?

**F** $\frac{4}{15}$ ft
**H** $1\frac{2}{3}$ ft
**G** $\frac{3}{5}$ ft
**J** $3\frac{3}{4}$ ft

Name ______________________ Date __________ Class __________

LESSON 12-1

# Reading Strategies
## *Analyzing Relationships*

The diagram below explains what inverse variation is, how to identify it, and how to answer questions involving it.

An *inverse variation* is a relationship that can be written using the formula $y = \frac{k}{x}$, where $x \neq 0$ and $k$ is the *constant of variation.*

The relationship shown in the table

| $x$ | −2 | 1 | 3 |
|---|---|---|---|
| $y$ | −3 | 6 | 2 |

is an inverse variation because $k$, or $x$ times $y$, is the same (6) for each ordered pair.

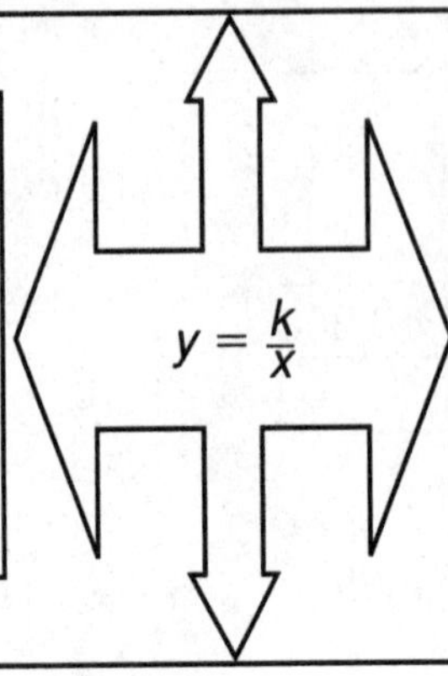


The equation $y = \frac{4}{x}$ is an inverse variation because it is in the form $y = \frac{k}{x}$, where $k = 4$.

If the value of $y$ varies inversely as $x$ and $y_1 = 3$ when $x_1 = 4$, we can find $y_2$ when $x_2 = 6$ by solving $x_1y_1 = x_2y_2$.

$$x_1y_1 = x_2y_2$$
$$(4)(3) = 6y_2$$
$$12 = 6y_2$$
$$2 = y_2$$

**Identify *k*, the constant of variation, for each inverse variation.**

1. $y = \frac{5}{x}$ ____________

2. $y = -\frac{1}{x}$ ____________

3. $xy = -7$ ____________

**Tell whether or not each relationship is an inverse variation.**

4. $y + 9 = x$ ____________

5. $y = \frac{3}{x}$ ____________

6.

| $x$ | −4 | −2 | −1 |
|---|---|---|---|
| $y$ | −2 | −4 | −8 |

____________

7.

| $x$ | 1 | 2 | 3 |
|---|---|---|---|
| $y$ | 3 | 2 | 1 |

____________

**Let *y* vary inversely as *x*. Find each indicated value.**

8. $x_1 = 1$, $y_1 = -9$, $y_2 = 3$; find $x_2$ ____________

9. $x_1 = 4$, $y_1 = 5$, $x_2 = -2$; find $y_2$ ____________

10. $x_1 = 3$, $y_1 = 5$, $y_2 = 1$; find $x_2$ ____________

11. $x_1 = 2$, $y_1 = -4$, $x_2 = -4$; find $y^2$ ____________

Name ______________________ Date ____________ Class ____________

LESSON 12-2

# Practice A
## *Rational Functions*

**Identify the excluded value for each rational function.**

**1.** $y = \frac{2}{x}$ ____________

**2.** $y = \frac{2}{x + 1}$ ____________

**3.** $y = \frac{5}{x - 7}$ ____________

**Identify the asymptotes.**

**4.** $y = \frac{1}{x - 6}$ ____________

**5.** $y = \frac{2}{x + 4} - 5$ ____________

**6.** $y = \frac{3}{x - 1} + 4$ ____________

**Graph each function.**

**7.** $y = \frac{8}{x + 2}$

**a.** Vertical asymptote: ____________

**b.** Horizontal asymptote: ____________

**c.** Graph.

| *x* | *y* |
|---|---|
| −10 | |
| −6 | |
| −4 | |
| 0 | |
| 2 | |
| 6 | |

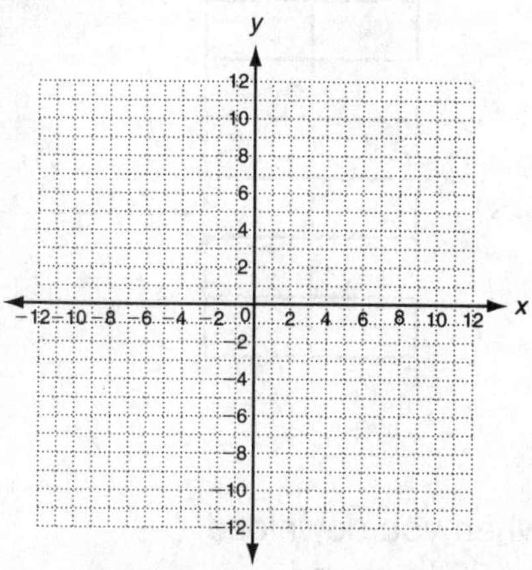


**8.** $y = \frac{20}{x + 5} - 5$

**a.** Vertical asymptote: ____________

**b.** Horizontal asymptote: ____________

**c.** Graph.

| *x* | *y* |
|---|---|
| −25 | |
| −15 | |
| −10 | |
| 0 | |
| 5 | |
| 15 | |

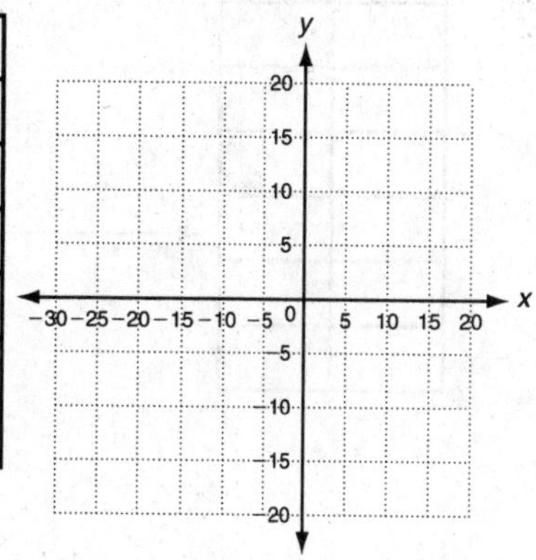


**9.** A caterer has $50 in his budget for cheddar cheese. If *x* represents the cost per pound, then $y = \frac{50}{x}$ represents the number of pounds *y* he can buy.

**a.** Describe the reasonable domain and range values.

______________________________

______________________________

**b.** Graph the function.

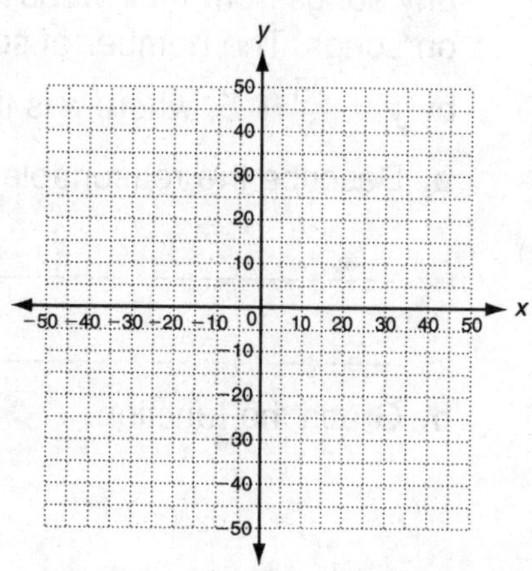

Name ______________________ Date ____________ Class ____________

**LESSON 12-2**

# Practice B

## *Rational Functions*

**Identify the excluded value for each rational function.**

**1.** $y = \frac{-6}{x}$ ______________

**2.** $y = \frac{8}{x + 3}$ ______________

**3.** $y = \frac{5}{3x - 6}$ ______________

**Identify the asymptotes.**

**4.** $y = \frac{5}{2x - 5}$ ______________

**5.** $y = \frac{2}{x + 3} - 4$ ______________

**6.** $y = \frac{4}{x - 6} + 3$ ______________

**Graph each function.**

**7.** $y = \frac{12}{x + 3}$

**a.** Vertical asymptote: ______________

**b.** Horizontal asymptote: ______________

**c.** Graph.

| *x* | *y* |
|---|---|
| | |
| | |
| | |
| | |
| | |
| | |

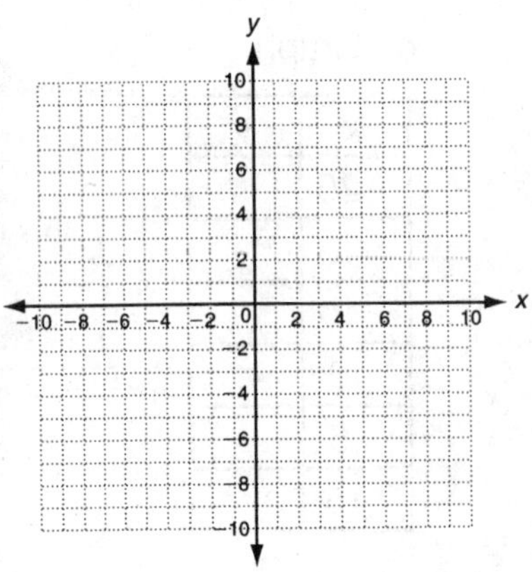


**8.** $y = \frac{6}{x - 1} - 3$

**a.** Vertical asymptote: ______________

**b.** Horizontal asymptote: ______________

**c.** Graph.

| *x* | *y* |
|---|---|
| | |
| | |
| | |
| | |
| | |
| | |

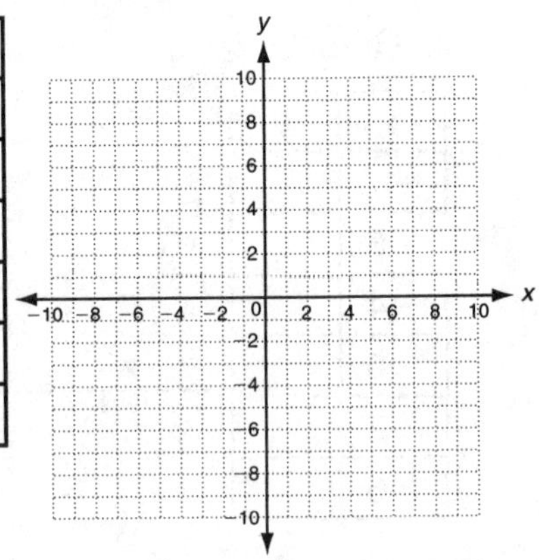


**9.** A music website is offering 5 free songs when you download any songs from their website. Catrina has \$25 to spend on songs. The number of songs $y$ that she can buy is given by $y = \frac{25}{x} + 5$, where $x$ is the price per song.

**a.** Describe the reasonable domain and range values.

______________________________

______________________________

**b.** Graph the function.

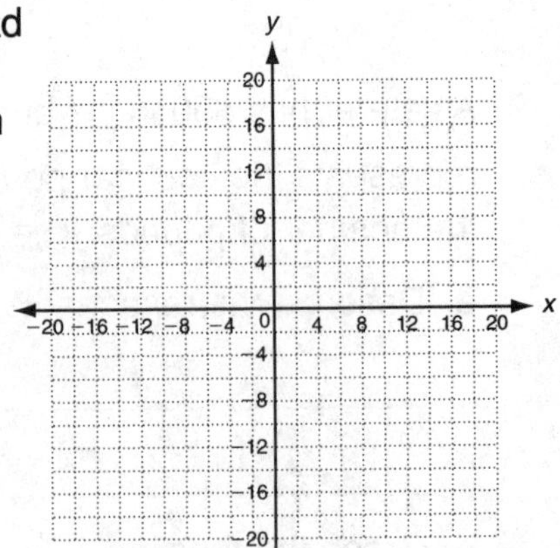

Name ______________________ Date __________ Class __________

LESSON 12-2

# Practice C

## Rational Functions

**Identify the excluded value for each rational function.**

1. $y = \frac{4}{3x}$ ______________

2. $y = \frac{2}{4x + 3}$ ______________

3. $y = \frac{5}{6x - 1}$ ______________

**Identify the asymptotes.**

4. $y = \frac{1}{x - 0.8}$ ______________

5. $y = \frac{6}{5x - 1} + 2$ ______________

6. $y = \frac{2}{0.5x + 3} - 7$ ______________

**Graph each function.**

7. $y = \frac{3}{x - 4}$

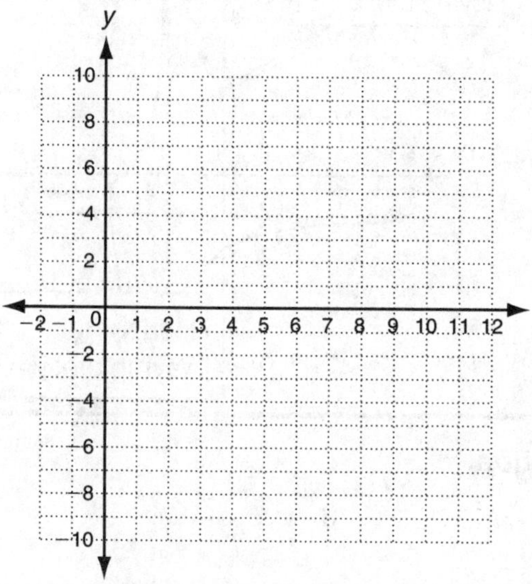


8. $y = \frac{10}{x + 2} - 4$

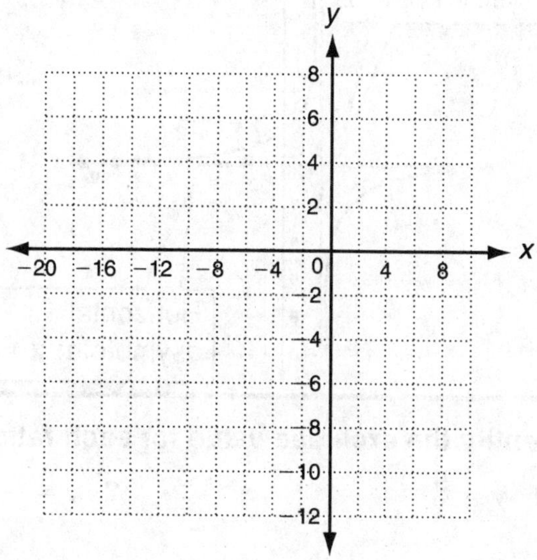


9. A furniture store is offering a payment plan which includes a finance charge of \$15 per month. The time in months $y$ it will take to pay of a bill of \$450, when $x$ dollars are paid each month, is $y = \frac{450}{x - 15}$.

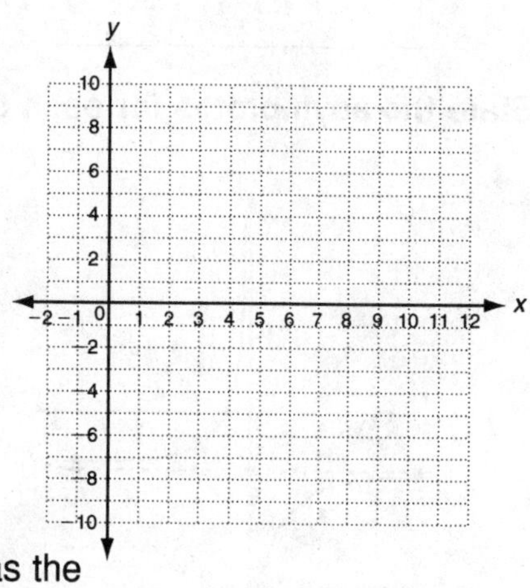


   a. Describe the reasonable domain and range values.

   ______________________________

   ______________________________

   b. Graph the function.

10. Write a rational function whose graph is the same shape as the graph of $f(x) = \frac{1}{x}$, but is translated 3 units right and 6 units up. ______________

Name ______________________ Date __________ Class __________

LESSON 12-2

# Reteach

## Rational Functions

Remember that division by zero is undefined. Because rational functions have $x$ in the denominator, we must exclude any values that make the denominator equal to zero.

**Identify the excluded value for $y = \dfrac{3}{x - 2}$.**

The function will be undefined when $x - 2 = 0$.

$$\begin{aligned} x - 2 &= 0 \\ +2 \quad &\; +2 \\ x &= 2 \end{aligned}$$

The excluded value is 2.

For rational functions, a vertical asymptote will occur at excluded values. An asymptote is a line that a graph gets close to, but never touches. Most rational functions have a vertical and a horizontal asymptote.

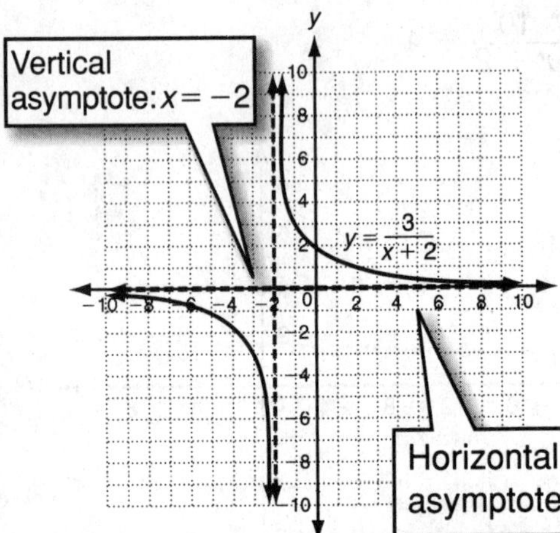


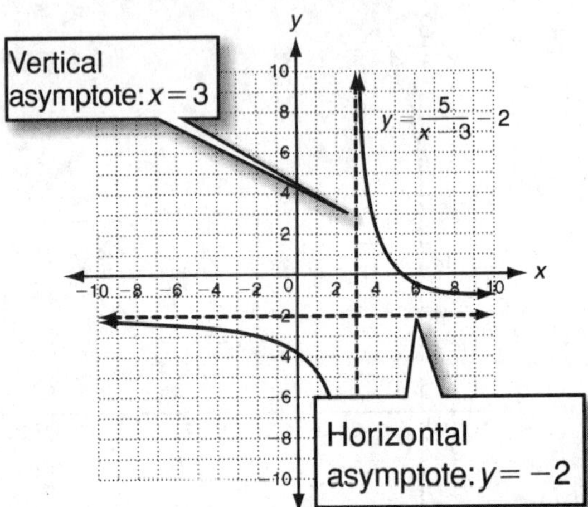


**Identify the excluded value for each rational function.**

**1.** $y = \dfrac{5}{x}$ ______________

**2.** $y = -\dfrac{6}{x - 5}$ ______________

**3.** $y = \dfrac{4}{x + 7}$ ______________

**State the asymptotes for each graph below.**

**4.**

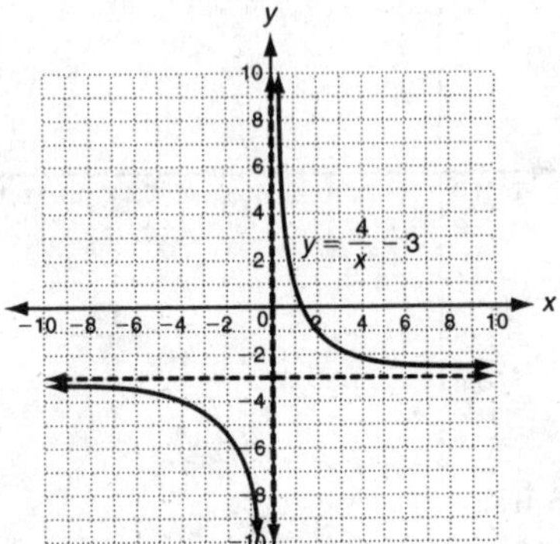


______________

**5.**

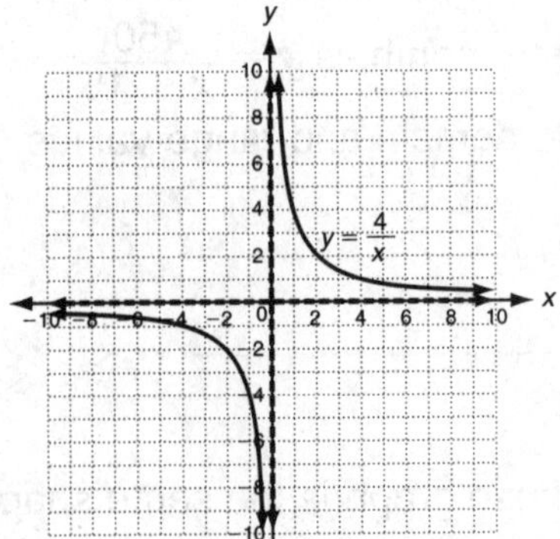


______________

**6.**

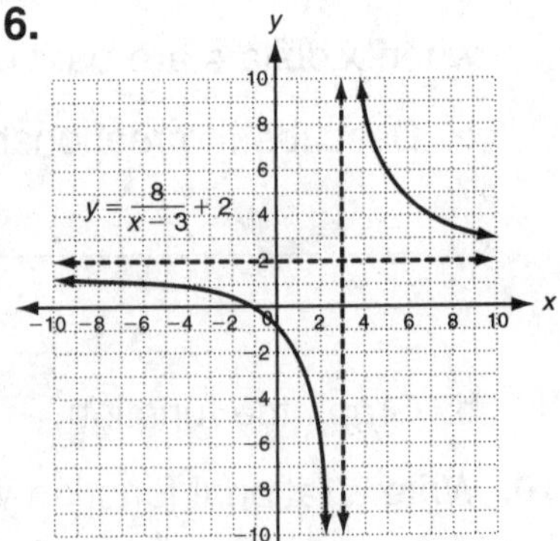


______________

Name ______________________ Date __________ Class __________

LESSON 12-2

# Reteach

## *Rational Functions (continued)*

You can also determine the asymptotes from the rational function itself.

$$y = \frac{a}{x + b} + c$$

- vertical asymptote at $x = -b$
- horizontal asymptote at $y = c$

You can use the asymptotes to help you graph.

**Graph $y = \frac{12}{x + 2}$.**

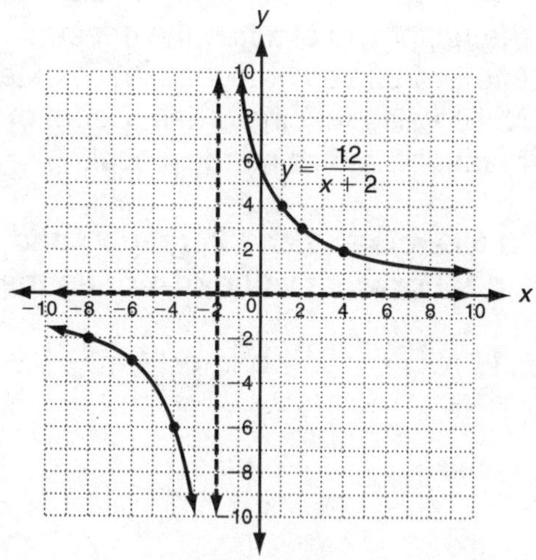


**Step 1:** Identify the vertical and horizontal asymptotes.

$b = 2$, so the vertical asymptote is $x = -2$.

$c = 0$, so the horizontal asymptote is $y = 0$.

**Step 2:** Graph the asymptotes.

**Step 3:** Make a table of values.

Choose $x$ values on both sides of the vertical asymptote.

| $x$ | −8 | −6 | −4 | 1 | 2 | 4 |
|---|---|---|---|---|---|---|
| $y$ | −2 | −3 | −6 | 4 | 3 | 2 |

**Identify the asymptotes.**

7. $y = \frac{4}{x + 5}$ ______________

8. $y = \frac{3}{x + 2} - 6$ ______________

9. $y = \frac{1}{x - 3} + 8$ ______________

**Graph each function.**

10. $y = \frac{4}{x - 1}$

asymptotes: ______________

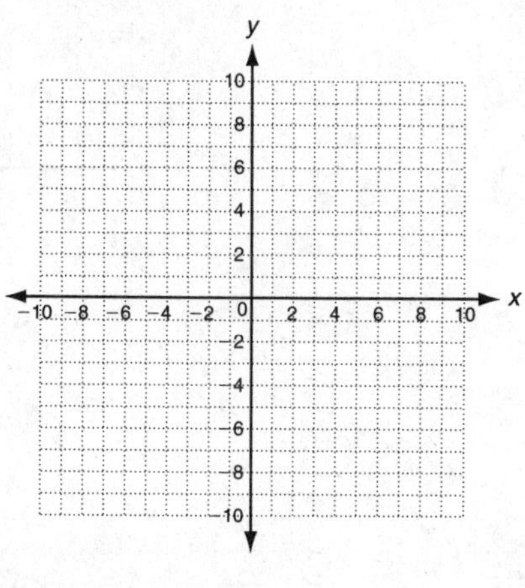


11. $y = \frac{8}{x + 2} + 3$

asymptotes: ______________

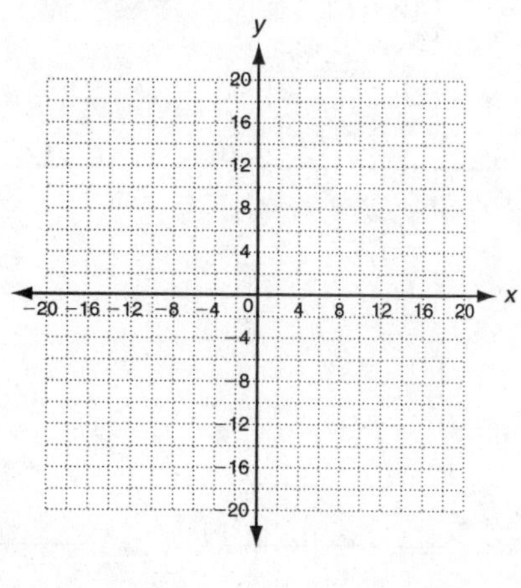

Name ______________________ Date __________ Class __________

LESSON 12-2

# Challenge

## Exploring Functions Defined by Reciprocals

The function $f(x) = x$ is a polynomial function. The function $g(x) = \frac{1}{x}$ is a rational function. The diagram at right shows the graphs of $f(x) = x$ and $g(x) = \frac{1}{x}$.

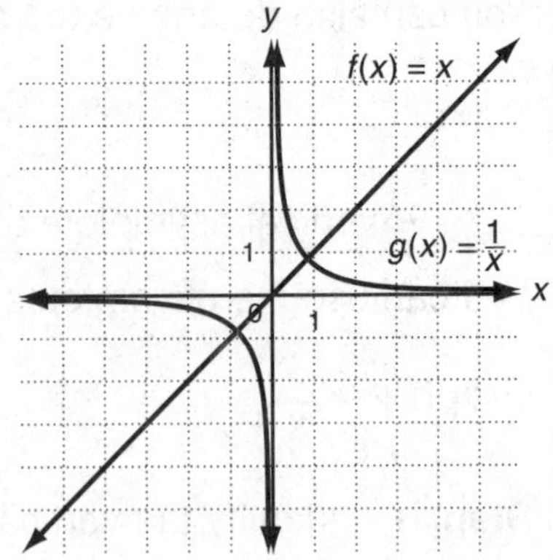


Notice that the graph of $f$ is a continuous and straight line. The graph of $g$ is neither continuous nor straight. Taking the reciprocal of $x$ has the effect of breaking a straight line into two curved branches, each extending indefinitely. Notice also that the break occurs at $x = 0$, the only real number that has no reciprocal.

**In Exercises 1 and 2, graph $f$ and $g$ on the same coordinate grid. For what value of $x$ does the break occur?**

1. $f(x) = x + 1$ and $g(x) = \frac{1}{x + 1}$

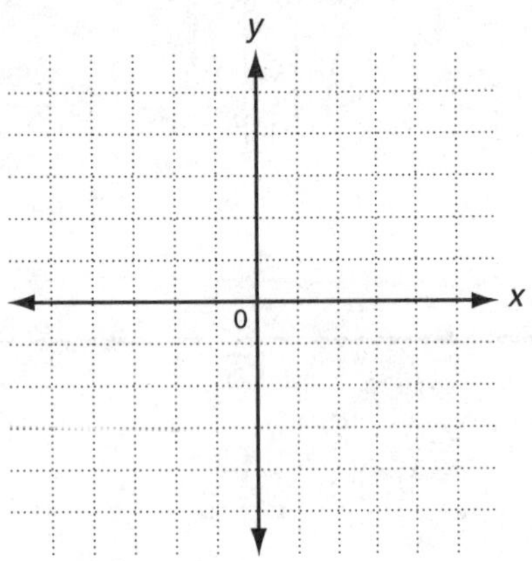


______________________

2. $f(x) = x - 1$ and $g(x) = \frac{1}{x - 1}$

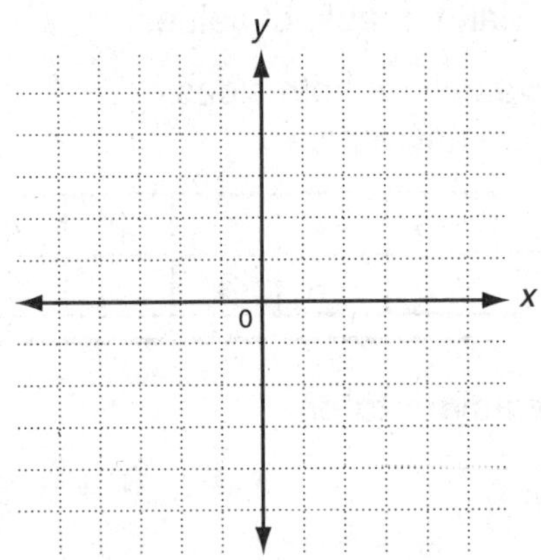


______________________

**Graph each function and its reciprocal on the same coordinate grid.**

3. $y = x - 2.5$

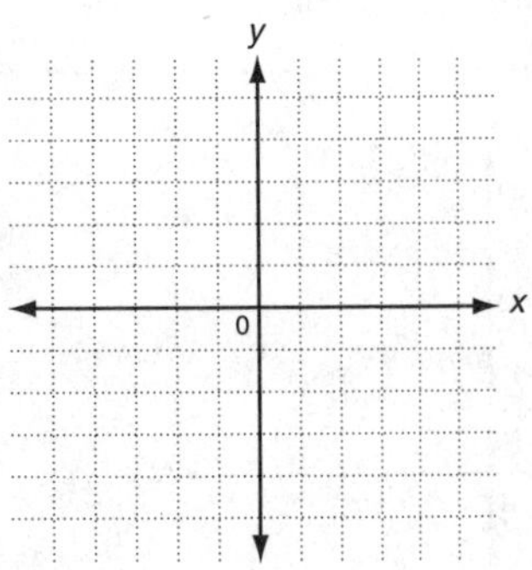


______________________

4. $y = x - 3$

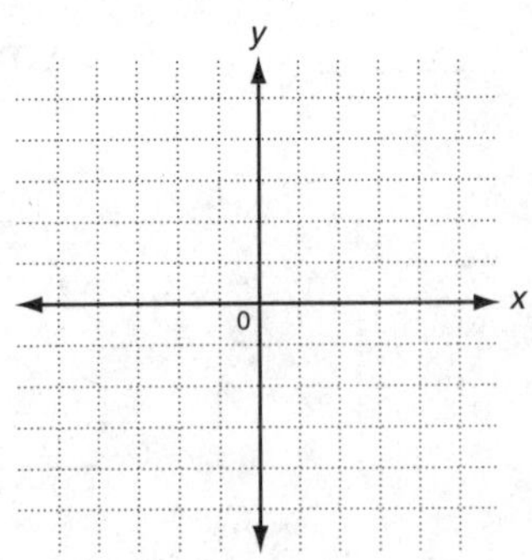


______________________

5. $y = x + 3$

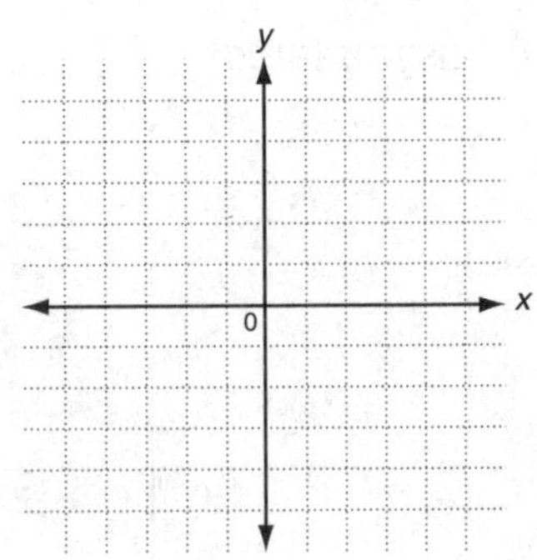


______________________

Name ______________________ Date __________ Class __________

LESSON 12-2

# Problem Solving

## *Rational Functions*

**Gabrielle saves \$100 and decides to join a CD Club. As a member, she gets 12 CDs free and then must buy the rest at the regular price of $x$ dollars. The number of CDs $y$ she can buy is $y = \frac{100}{x} + 12$.**

**1.** Determine a reasonable domain and range.

______________________

______________________

**2.** Identify the vertical and horizontal asymptotes.

______________________

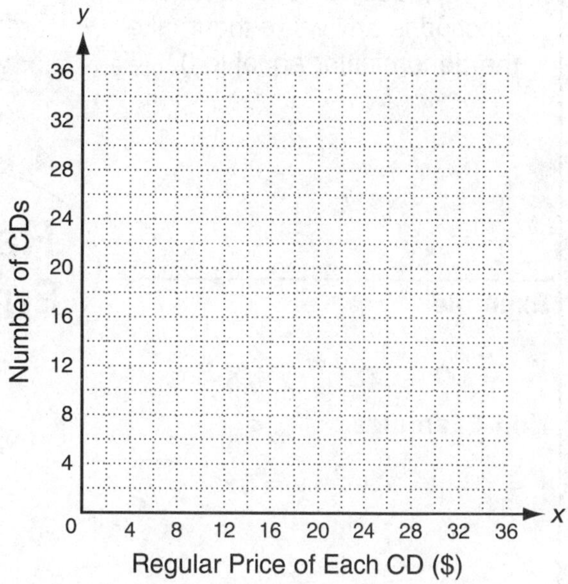


**3.** Complete this table of ordered pairs and graph the rational function.

| *x* | 5 | 10 | 20 | 25 |
|---|---|---|---|---|
| *y* | | | | |

**4.** Gabrielle learns that the regular price of a CD is \$17.90. How many CDs can she buy through the Club with \$100?

______________________

**Select the best answer. For 5–6, refer to the situation above.**

**5.** Before joining the CD Club, Gabrielle learns about the Student Savings Card. If she buys a card for \$20, she gets \$5 off every CD she buys at the local record store. If each CD costs $x$ dollars at the local record store, then the number of CDs $y$ she can buy is $y = \frac{100 - 20}{x - 5}$. The average price of each CD at the local record store is \$14.95. How many CDs can she buy with the Card and \$100?

**A** 4 **C** 10

**B** 8 **D** 12

**6.** Gabrielle realizes she must pay \$2.95 for shipping and handling on every CD from the club, including the 12 free CDs. That changes the function for the club to $y = \frac{100 - 2.95(12)}{x + 2.95} + 12$. If the regular price is still \$17.90, how many CDs can she buy through the Club?

**F** 3 **H** 16

**G** 15 **J** 18

**7.** At $x$ meters away from a very loud stereo speaker, the intensity of the sound $y$ in Watts per square meter is given by $y = \frac{0.0016}{x^2}$. What is the intensity when you stand 4 meters away?

**A** 0.0001 W/m$^2$ **C** 0.0004 W/m$^2$

**B** 0.0002 W/m$^2$ **D** 0.0008 W/m$^2$

Name ______________________ Date __________ Class __________

LESSON 12-2

# Reading Strategies

## *Use a Concept Map*

The concept map below will help you understand rational functions.

**Definitions**

A *rational function* is a function whose rule has a variable in the denominator.

An *excluded value* for a rational function is any value that makes the denominator equal to 0.

**Graph**

$y = \frac{1}{x-1} + 2$

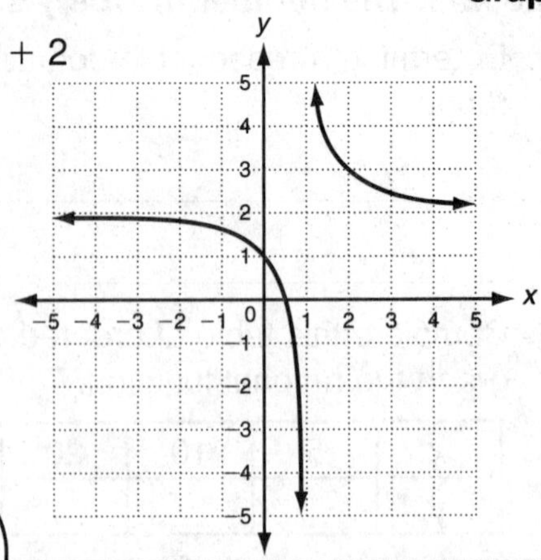


**Rational Function**

**Examples**

$y = \frac{4}{x}$; $y = \frac{1}{x^2}$; $y = \frac{2}{x-5} + 3$

**Non-Examples**

$y = \frac{x}{6}$; $y = x + \frac{1}{2}$; $y = 3x^2$

**Asymptotes**

A rational function in the form $y = \frac{a}{x+b} + c$ has a *vertical asymptote* at the excluded value $x = -b$, and a *horizontal asymptote* at $y = c$.

**Complete the following.**

1. Tell whether or not each of the following is a rational function.

**a.** $y = \frac{x}{3} + 2$ **b.** $y = \frac{3}{x^2} - 1$ **c.** $y = \frac{1}{x+8}$

______________ ______________ ______________

2. Explain why $y = x + \frac{1}{2}$ is not a rational function. ______________________

______________________________________________

3. Describe how to find the excluded value of the rational function $y = \frac{3}{x-5}$. Then find it.

______________________________________________

4. Identify the asymptotes for the rational function $y = \frac{1}{x-1} + 2$ graphed above.

vertical: ______________

horizontal: ______________

Name ______________________ Date ____________ Class ____________

LESSON 12-3

# Practice A

## Simplifying Rational Expressions

**Find any excluded values for each rational expression.**

**1.** $\frac{2}{x-5}$ ____________

**2.** $\frac{4}{x^2+3x}$ ____________

**3.** $\frac{x+3}{x^2+6x+8}$ ____________

**Simplify each rational expression, if possible.**

**4.** $\frac{x+4}{x^2-16}$

$\frac{x+4}{(\square)(\square)}$

____________

**5.** $\frac{3x-18}{x^2-36}$

$\frac{3(\square)}{(\square)(\square)}$

____________

**6.** $\frac{x^2-1}{x^2+8x+7}$

$\frac{(\square)(\square)}{(\square)(\square)}$

____________

**7.** $\frac{x+5}{x^2+7x+10}$ ____________

**8.** $\frac{x^2+4x-5}{x^2-4x+3}$ ____________

**9.** $\frac{x^2-2x}{x^2+2x-8}$ ____________

**10.** $\frac{3-x}{2x-6}$

$\frac{-1(\square)}{2(\square)}$

____________

**11.** $\frac{x-4}{16-x^2}$ ____________

**12.** $\frac{8-x}{x^2-6x-16}$ ____________

**13.** An animal's surface-area-to-volume ratio indicates the amount of heat loss the animal may experience. Because a smaller animal has a higher ratio of surface-area-to-volume, the smaller animal has a higher rate of heat loss.

**a.** Find the surface-area-to-volume ratio of a cube.
(*Hint*: For a cube, $S = 6s$ and $V = s^3$.) ____________

**b.** Two animals are both close to a cube in shape. The height of the first animal is 4 cm. The height of the second animal is 6 cm. Which animal would experience the greater amount of heat loss? Explain.

________________________________________

________________________________________

Name ____________________ Date __________ Class __________

LESSON 12-3

# Practice B

## Simplifying Rational Expressions

**Find any excluded values for each rational expression.**

1. $\dfrac{6}{3 + x}$ ____________

2. $\dfrac{5}{x^2 - 4x}$ ____________

3. $\dfrac{x + 6}{x^2 + 3x - 4}$ ____________

**Simplify each rational expression, if possible. Identify any excluded values.**

4. $\dfrac{7}{x - 3}$ ____________

5. $\dfrac{5x^2 + 10x}{5x}$ ____________

6. $\dfrac{2x}{4x^2 + 6x}$ ____________

**Simplify each rational expression, if possible.**

7. $\dfrac{x + 3}{x^2 - 2x - 15}$ ____________

8. $\dfrac{3x + 6}{x^2 + 3x + 2}$ ____________

9. $\dfrac{x - 6}{x^2 - 7x + 6}$ ____________

10. $\dfrac{x^2 - 49}{x^2 + 8x + 7}$ ____________

11. $\dfrac{x^2 + 4x - 5}{x^2 - 4x + 3}$ ____________

12. $\dfrac{x^2 - 2x}{x^2 + 2x - 8}$ ____________

13. $\dfrac{x^2 - x - 12}{4 - x}$ ____________

14. $\dfrac{5 - 5x}{x^2 - 1}$ ____________

15. $\dfrac{3 - x}{x^2 - 6x + 9}$ ____________

16. When packaging food, a company wants a package that uses the least amount of material to hold the greatest volume of product. Some containers with mixed nuts are in the shape of a right circular cylinder.

    **a.** Find the surface-area-to-volume ratio of a right circular cylinder. (Hint: For a right circular cylinder, $S = 2\pi rh + 2\pi r^2$ and $V = \pi r^2 h$.) ____________

    **b.** Container A has a radius of 2 in. and a height of 5 in. Container B has a radius of 4 in. and a height of 8 in. Which container should the company choose? Explain.

    ______________________________

    ______________________________

Name ______________________ Date __________ Class __________

LESSON 12-3

# Practice C

## *Simplifying Rational Expressions*

**Find any excluded values for each rational expression.**

1. $\dfrac{5}{4x - 7}$

2. $\dfrac{4x}{x^2 - 4x}$

3. $\dfrac{x - 5}{x^2 - 25}$

**Simplify each rational expression, if possible. Identify any excluded values.**

4. $\dfrac{x - 1}{x^2 - 4}$

5. $\dfrac{4x^2 - 20x}{2x - 10}$

6. $\dfrac{x + 1}{2x^3 + 5x^2 + 3x}$

**Simplify each rational expression, if possible.**

7. $\dfrac{x^2 + 4x - 5}{x^2 + 8x + 15}$

8. $\dfrac{3x - 12}{6x^2 - 18x - 24}$

9. $\dfrac{2x^2 - 32}{x^2 + 8x + 16}$

10. $\dfrac{10x^2 - 360}{5x^2 + 30x}$

11. $\dfrac{x^2 + 2x - 8}{x^3 - 2x^2 - 24x}$

12. $\dfrac{x^2 - 1}{2x^2 - 3x - 5}$

13. $\dfrac{5 - x}{2x^3 - 16x^2 + 30x}$

14. $\dfrac{x^2 - 25}{10 + 3x - x^2}$

15. $\dfrac{27 - 3x^2}{6x^2 - 36x + 54}$

16. Flavored crushed ice is served in a paper container in the shape of a right circular cone.

 a. Find the surface-area-to-volume ratio of a right circular cone. (*Hint*: For a right circular cone, $S = \pi r^2 + \pi rs$ (where $s = \sqrt{r^2 + h^2}$) and $V = \frac{1}{3}\pi r^2 h$.)

 b. Container A has a radius of 2 in. and a height of 3 in. Container B has a radius of 3 in. and a height of 4 in. Which container uses the least amount of paper to hold the greatest amount of crushed ice? Explain.

Name ______________________ Date __________ Class __________

LESSON 12-3

# Reteach

## Simplifying Rational Expressions

A **rational expression** is an algebraic expression whose numerator and denominator are polynomials.

Exclude any values from a rational expression that make the denominator equal zero.

**Find any excluded value of $\frac{6x}{x^2 - 5x}$.**

$\frac{6x}{x^2 - 5x}$

$x^2 - 5x = 0$ *Set denominator = 0.*

$x(x - 5) = 0$ *Factor.*

$x = 0$ or $x - 5 = 0$ *Zero Product*

$+5 \quad +5$ *Property*

$x = 5$

**The excluded values are 0 and 5.**

**Simplify $\frac{x + 4}{2x^2 + 8x}$, if possible. Identify any excluded values.**

$\frac{x + 4}{2x^2 + 8x}$

$\frac{x + 4}{2x(x + 4)}$ Factor the denominator.

Find excluded values here.

$\frac{\cancel{x + 4}^{1}}{2x\,(\cancel{x + 4})^{1}}$ Divide out common factors.

$\frac{1}{2x}$ Simplify.

Remember to find the excluded value from the original equation (not the simplified one). **The excluded values are 0 and −4.**

**Identify any excluded values.**

1. $\frac{3}{4x}$ ______________

2. $\frac{5x}{3x^2 + 15x}$ ______________

3. $\frac{6}{x^2 - 5x - 14}$ ______________

**Simplify each rational expression, if possible. Identify any excluded values.**

4. $\frac{4x}{20x^3}$ ______________

5. $\frac{x + 3}{x - 4}$ ______________

6. $\frac{27x}{3x^2}$ ______________

7. $\frac{6}{3x + 9}$ ______________

8. $\frac{x + 4}{2x^2 + 8x}$ ______________

9. $\frac{5x^2 - 20x}{x - 4}$ ______________

Name ______________________ Date ____________ Class ____________

LESSON 12-3

# Reteach

## Simplifying Rational Expressions (continued)

Rational expressions with binomials, trinomials and opposite binomials can also be simplified.

**Simplify $\frac{x^2 - 4}{x^2 + 9x + 14}$, if possible.**

$\frac{x^2 - 4}{x^2 + 9x + 14}$

$\frac{(x + 2)(x - 2)}{(x + 2)(x + 7)}$ *Factor.*

$\frac{\cancel{(x + 2)}^1(x - 2)}{\cancel{(x + 2)}^1(x + 7)}$ *Divide out the common factors.*

$\frac{x - 2}{x + 7}$ *Simplify.*

**Simplify $\frac{3 - x}{x^2 - 7x + 12}$, if possible.**

$\frac{3 - x}{x^2 - 7x + 12}$

$\frac{3 - x}{(x - 3)(x - 4)}$ *Factor.*

$\frac{-1(x - 3)}{(x - 3)(x - 4)}$ *Factor −1 out of the numerator.*

$\frac{-1\cancel{(x - 3)}^1}{\cancel{(x - 3)}^1(x - 4)}$ *Divide out the common factors.*

$-\frac{1}{x - 4}$ *Simplify.*

**Simplify each rational expression, if possible.**

10. $\frac{x + 3}{x^2 - 9}$

$\frac{(\square)}{(\square)(\square)}$

____________

11. $\frac{x^2 - 4x}{x^2 - x - 12}$

$\frac{\square(\square)}{(\square)(\square)}$

____________

12. $\frac{x^2 - 2x - 15}{x^2 + 5x + 6}$

$\frac{(\square)(\square)}{(\square)(\square)}$

____________

13. $\frac{x - 9}{x^2 - 81}$

____________

14. $\frac{x^2 + 7x + 6}{x + 1}$

____________

15. $\frac{x + 3}{x^2 + 7x + 12}$

____________

16. $\frac{3x + 15}{x^2 - 25}$

____________

17. $\frac{x - 5}{25 - x^2}$

____________

18. $\frac{4 - x}{x^2 - 2x - 8}$

____________

Name ______________________ Date ____________ Class ____________

LESSON 12-3

# Challenge

## More About Graphing Rational Functions

A rational function in the form $y = \frac{P(x)}{Q(x)}$, where $P(x)$ and $Q(x)$ are polynomials, may have vertical asymptotes, holes, or horizontal asymptotes according to the following rules.

**Example:**

$$y = \frac{x-1}{(x-1)(x-2)}$$

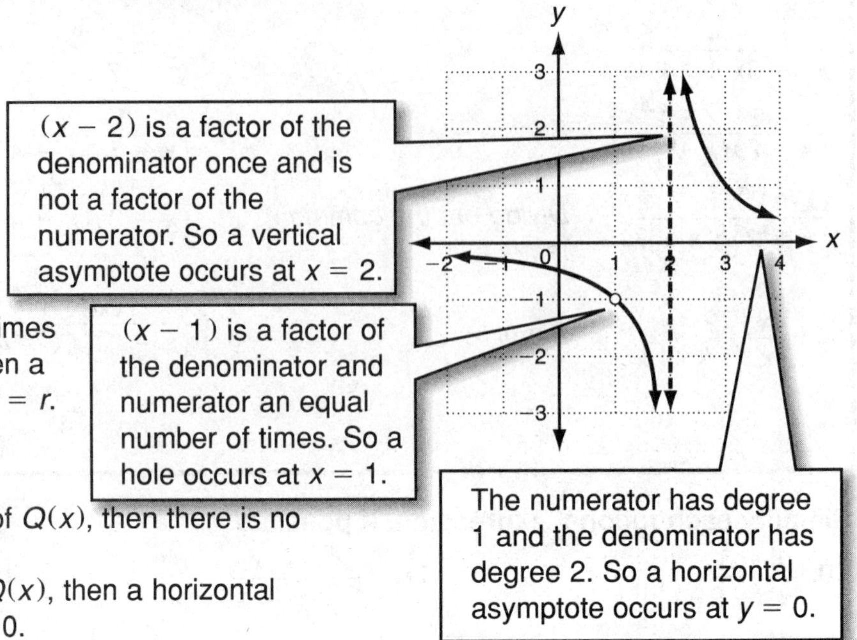


**Vertical Asymptotes**

If $(x - r)$ is a factor of $Q(x)$ a greater numberof times than it is a factor of $P(x)$, then a vertical asymptote occurs at $x = r$.

**Holes**

If $(x - r)$ is a factor of $Q(x)$ a fewer or equal number of times than it is a factor of $P(x)$, then a hole occurs in the graph at $x = r$.

**Horizontal Asymptotes**

If the degree of $P(x)$ is ...

- greater than the degree of $Q(x)$, then there is no horizontal asymptote.
- less than the degree of $Q(x)$, then a horizontal asymptote occurs at $y = 0$.
- equal to the degree of $Q(x)$, then a horizontal asymptote occurs at $y = \frac{\text{leading coefficient of } P(x)}{\text{leading coefficient of } Q(x)}$.

**Graph these rational functions. First factor the numerator and denominator. Then use the rules above to identify asymptotes and holes. Finally, use a table of values to fill in-between.**

**1.** $y = \frac{3x^2 - 7x - 6}{x^2 - 2x - 3} =$ ____________

asymptotes: ____________

holes: ____________

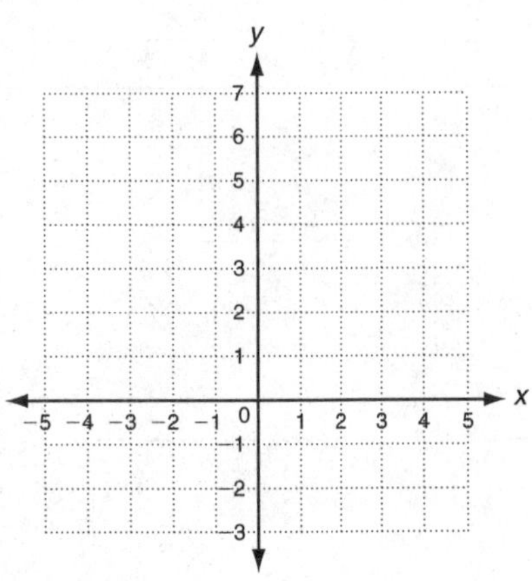

**2.** $y = \frac{x^2 - 3x - 4}{-x^2 + 4} =$ ____________

asymptotes: ____________

holes: ____________

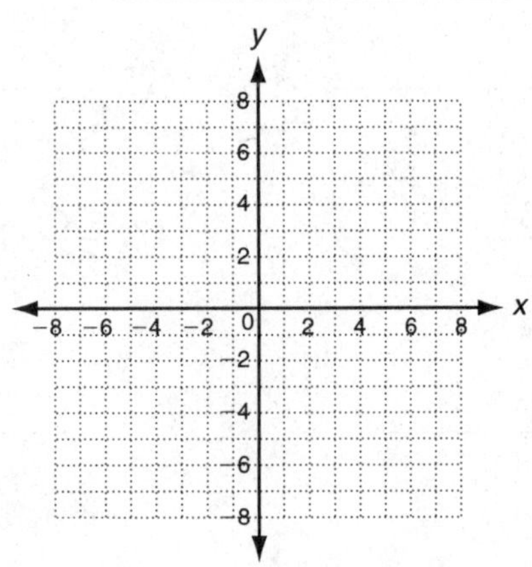

Name ______________________ Date ____________ Class ____________

LESSON 12-3

# Problem Solving

## *Simplifying Rational Expressions*

**A food company is considering two types of cylindrical cans for a new trail mix. The company wants a can that uses the least material for the greatest volume. For a cylinder, $S = 2\pi r^2 + 2\pi rh$ and $V = \pi r^2 h$.**

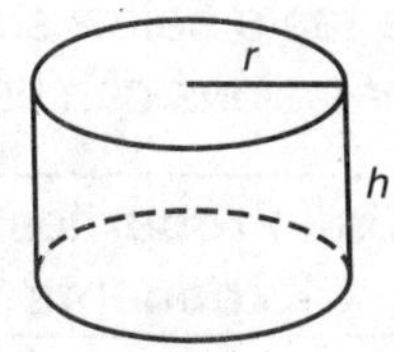


1. What is the surface-area-to-volume ratio for any cylinder?

_______________________

2. Can A has radius 2 in. and height 3 in. What is the surface-area-to-volume ratio for Can A?

_______________________

3. Can B has radius 1.5 in. and height 4 in. What is the surface-area-to-volume ratio for Can B?

_______________________

4. Which can should the company choose? Explain.

_______________________

_______________________

_______________________

**Select the best answer.**

5. For any circle with radius $r$, $P = 2\pi r$ and $A = \pi r^2$. Which expression represents the perimeter-to-area ratio?

**A** $2r$ **C** $\frac{r}{2}$

**B** $\frac{2}{r}$ **D** $\frac{1}{2r}$

6. Tyrone takes an 8-by-8 square piece of cardboard and cuts $x$-by-$x$ squares out of each corner. (All units are inches.) He then folds up the sides to form a box with no top. Which expression represents the surface-area-to-volume ratio? (*Hint:* For the surface area, find the area of the entire piece of cardboard and subtract the squares that were removed.)

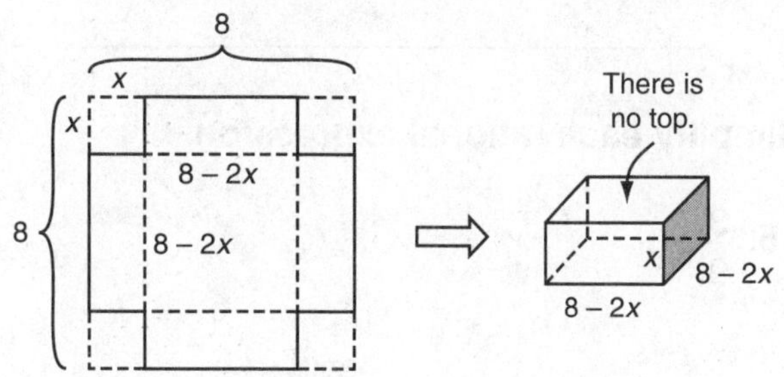


**F** $\frac{1}{x}$ **H** $\frac{8}{x(4 - x)}$

**G** $\frac{1}{2x(1 - 16x)}$ **J** $\frac{4 + x}{x(4 - x)}$

7. In 2000, there are $x$ mountain lions (predators) and $x$ deer (prey) in a wildlife preserve. In 2005, the populations level off at $3x$ mountain lions and $4x$ deer, for a predator-to-prey ratio of $\frac{3}{4}$. If the wildlife preserve adds 12 more deer in 2007, how many mountain lions should also be added to keep the predator-to-prey ratio equivalent to $\frac{3}{4}$?

**A** 9 **C** 16

**B** 12 **D** 20

Name ______________________ Date ____________ Class ____________

LESSON 12-3

# Reading Strategies
## *Use a Table*

The table below shows examples of the many ways that rational expressions can be simplified.

| Use Properties of Exponents | Factor out Common Monomials | Factor Trinomials & Special Products | Use Opposite Binomials |
|---|---|---|---|
| $\frac{4m^5}{24m^2}$ | $\frac{2b^2 - 18b}{b - 9}$ | $\frac{x^2 - 9}{x^2 + 2x - 15}$ | $\frac{6t - t^2}{t^2 - 4t - 12}$ |
| $\frac{m^3}{6}$ | $\frac{2b(b - 9)}{b - 9}$ | $\frac{(x + 3)(x - 3)}{(x + 5)(x - 3)}$ | $\frac{t(6 - t)}{(t + 2)(t - 6)}$ |
| | $2b$ | $\frac{x + 3}{x + 5}$ | $\frac{t(-1)(t - 6)}{(t + 2)(t - 6)}$ |
| | | | $-\frac{t}{(t + 2)}$ |

**Use the table to answer the following.**

1. Look at the first column. Explain how $\frac{4m^5}{24m^2}$ was simplified to $\frac{m^3}{6}$. ____________

____________________________________________

2. Look at the second column. What expression was divided out of the numerator and denominator? ____________________________

3. Look at the third column. What type of special product was in the numerator?

____________________________________________

4. Look at the fourth column. What happened in the numerator?

____________________________________________

**Simplify each rational expression.**

5. $\frac{3t^4}{9t^6}$

____________________

6. $\frac{n^4 - 3n^3}{2n^2 - 6n}$

____________________

7. $\frac{4c + 4}{2c^2 - 5c - 7}$

____________________

8. $\frac{18 - 2x}{x^2 - 81}$

____________________

Name ______________________ Date __________ Class __________

LESSON 12-4

# Practice A

## *Multiplying and Dividing Rational Expressions*

**Multiply. Simplify your answer.**

**1.** $\frac{3x^2}{2y^4} \cdot \frac{4xy^4}{x^6}$

**2.** $\frac{2q-8}{6q^2} \cdot \frac{q^5}{pq-4p}$

**3.** $\frac{ab}{5ab-5b} \cdot (a^2+3a-4)$

**4.** $\frac{2}{4m^2+8m} \cdot \frac{m^2+8m+12}{6+m}$

**Divide by multiplying the reciprocal. Simplify your answer.**

**5.** $\frac{c^4d}{9c} \div \frac{2cd^8}{3c^4d^2}$

**6.** $\frac{k^2-6k+9}{k+1} \div \frac{8k^2-24k}{2k^2}$

**7.** For a game at school, Alma is asked to pull the names of two students out of a bag without looking, and without replacing the first name. There are 3 more boys than girls in the class.

**a.** Let $x$ represent the number of girls in the class. Write an expression to represent the number of boys. __________

**b.** Write a simplified expression to represent the total number of students (girls + boys) in the class. __________

**c.** Write and simplify an expression that represents the probability of Alma picking a girl, then a boy. *Hint*: Use $P(A \text{ and } B) = P(A) \cdot P(B \text{ after } A)$. __________

**d.** Substitute 10 for $x$ in your expression from part c to find the probability that Alma picks a girl then a boy when there are 10 girls in the class. Round to the nearest whole percent. __________

Name ______________________ Date __________ Class __________

LESSON 12-4

# Practice B

## *Multiplying and Dividing Rational Expressions*

**Multiply. Simplify your answer.**

**1.** $\frac{8a^2b^5}{a^3} \cdot \frac{3a^2}{4b^9}$

**2.** $\frac{4x + 8}{3} \cdot \frac{6x}{x + 2}$

________________________

________________________

**3.** $\frac{7}{2t - 6} \cdot (t^2 + t - 12)$

**4.** $\frac{3x^2 + xy^3}{y^3} \cdot \frac{2xy + 8y}{4x + x^2}$

________________________

________________________

**Divide. Simplify your answer.**

**5.** $\frac{5j^2k^2}{3jk^5} \div \frac{10j^2k}{9j^3}$

**6.** $\frac{3c^2 + 24c}{c^2 - 2c + 1} \div \frac{c^2 + 9c + 8}{9c - 9}$

________________________

________________________

**7.** Ramon is playing a game in which he must pull two blocks out of a bag containing red and yellow blocks. He cannot look, and he cannot replace the block. The bag has 4 more red blocks than yellow blocks.

**a.** Write and simplify an expression that represents Ramon's probability of picking a red block, then a yellow block. ________________________

**b.** What is the probability that Ramon pulls a red block then a yellow block if there are 6 yellow blocks in the bag before his first pick? ________________________

**c.** What is the probability that Ramon pulls two yellow blocks if there are 6 yellow blocks in the bag before his first pick? ________________________

Name ______________________ Date __________ Class __________

LESSON 12-4

# Practice C

## *Multiplying and Dividing Rational Expressions*

**Multiply. Simplify your answer.**

**1.** $\frac{9rt^5}{5r^4} \cdot \frac{10r^2}{27t}$

______________________

**2.** $\frac{6a + 12}{3a + a^2} \cdot \frac{a + 3}{4a^2 + 8a}$

______________________

**3.** $\frac{2m}{5m + 20} \cdot (m^2 - 16)$

______________________

**4.** $\frac{p + 3}{p^2 - 25} \cdot \frac{3p - 15}{p^2 + p - 6}$

______________________

**Divide. Simplify your answer.**

**5.** $\frac{x^4y}{3z^5} \div \frac{x^2z^3}{9y^2z}$

______________________

**6.** $\frac{8n^2 - 8}{10n + 10n^2} \div (2n^3 + 6n^2 - 8n)$

______________________

**7.** At the town fair, Jillian is in charge of pulling two winning tickets out of a box containing tickets for a prize drawing. There are 12 more tickets from female entrants than from male entrants.

**a.** If the same person cannot win twice, write and simplify an expression that represents the probability of Jillian picking a female winner, then a male winner. ______________________

**b.** If the same person can win twice, write and simplify an expression that represents the probability of Jillian picking a female winner, then a female winner. ______________________

**c.** What is the probability that Jillian pulls two female winning tickets if the same person can win twice, and if there are 25 tickets from female entrants before her first pick? ______________________

Name ______________________ Date __________ Class __________

LESSON 12-4

# Reteach

## Multiplying and Dividing Rational Expressions

If $a$, $b$, $c$, and $d$ are nonzero polynomials, then $\frac{a}{b} \cdot \frac{c}{d} = \frac{ac}{bd}$.

You can make any expression rational by writing it with a denominator of 1.

**Multiply $\frac{x+1}{7} \cdot \frac{5}{6x+6}$. Simplify your answer.**

$\frac{x+1}{7} \cdot \frac{5}{6x+6}$

$\frac{5(x+1)}{7(6x+6)}$ *Multiply the numerators and the denominators.*

$\frac{5(x+1)}{7 \cdot 6(x+1)}$ *Factor.*

$\frac{5\cancel{(x+1)}^{1}}{7 \cdot 6\cancel{(x+1)}^{1}}$ *Simplify.*

$\frac{5}{42}$

**Multiply $(3x+12) \cdot \frac{2}{x^2 - x - 20}$. Simplify your answer.**

$(3x+12) \cdot \frac{2}{x^2 - x - 20}$

$\frac{3x+12}{1} \cdot \frac{2}{x^2 - x - 20}$ *Write as a rational expression.*

$\frac{3(x+4)}{1} \cdot \frac{2}{(x+4)(x-5)}$ *Factor.*

$\frac{3\cancel{(x+4)}^{1}}{1} \cdot \frac{2}{\cancel{(x+4)}^{1}(x-5)}$ *Simplify.*

$\frac{6}{x-5}$

**Multiply. Simplify your answer.**

1. $\frac{x+3}{5x} \cdot \frac{2}{4x+12}$

   $\frac{(\square)}{(\square)} \cdot \frac{(\square)}{(\square)(\square)}$

   ______________

2. $\frac{3x}{x+3} \cdot \frac{x^2+5x+6}{x}$

   ______________

3. $\frac{4x-12}{6x} \cdot \frac{x+3}{x^2-9}$

   ______________

4. $(4x+24)\frac{5}{x^2-36}$

   ______________

5. $(x+7)\frac{3x}{x^2+13x+42}$

   ______________

6. $(x^2-16)\frac{6}{x^2-x-12}$

   ______________

Name ______________________________ Date ____________ Class ____________

LESSON 12-4

# Reteach

## Multiplying and Dividing Rational Expressions (continued)

If $a$, $b$, $c$, and $d$ are nonzero polynomials, then $\frac{a}{b} \div \frac{c}{d} = \frac{a}{b} \cdot \frac{d}{c} = \frac{ad}{bc}$.

**Divide $\frac{3}{x+2} \div \frac{5}{x+2}$.**
**Simplify your answer.**

$\frac{3}{x+2} \div \frac{5}{x+2}$

$\frac{3}{x+2} \cdot \frac{x+2}{5}$ *Write as multiplication by the reciprocal.*

$\frac{3}{\cancel{x+2}^{1}} \cdot \frac{\cancel{x+2}^{1}}{5}$ *Divide out common factors.*

$\frac{3}{5}$ *Simplify.*

**Divide $\frac{2x-4}{x^2} \div (x-2)$.**
**Simplify your answer.**

$\frac{2x-4}{x^2} \div (x-2)$

$\frac{2x-4}{x^2} \div \frac{x-2}{1}$ *Write as a rational* expression.

$\frac{2x-4}{x^2} \cdot \frac{1}{x-2}$ *Write reciprocal.*

$\frac{2(x-2)}{x^2} \cdot \frac{1}{x-2}$ *Factor.*

$\frac{2(\cancel{x-2})^{1}}{x^2} \cdot \frac{1}{\cancel{x-2}^{1}}$ *Simplify.*

$\frac{2}{x^2}$

**Divide. Simplify your answer.**

**7.** $\frac{4x}{x^2} \div \frac{x^2}{7}$

$\frac{\square}{\square} \cdot \frac{\square}{\square}$

**8.** $\frac{8x^2y^3}{2xy} \div \frac{xy}{3}$

**9.** $\frac{x^2}{x-3} \div \frac{x+5}{x^2+2x-15}$

______________ ______________ ______________

**10.** $\frac{x+4}{x-5} \div (x^2 - x - 20)$

**11.** $\frac{x^2 - 4x - 5}{8} \div (x^2 + 2x + 1)$

**12.** $\frac{x+10}{x^2} \div (x^2 - 100)$

______________ ______________ ______________

Name ______________________ Date __________ Class __________

LESSON 12-4

# Challenge

## Multiplication and Division Puzzles (Revisited)

**For 1 and 2, draw a path from Start to Finish, multiplying by each new expression along the way, to find a product equal to the goal. Use only vertical or horizontal moves, and the path cannot cross itself.**

**1.** Goal: Product of 1

| | | |
|---|---|---|
| $\frac{x^2 - x - 6}{x + 1}$ **Start** | $\frac{x^2 + 2x}{x - 3}$ | $\frac{x^2 - 4}{x - 3}$ |
| $\frac{x}{x + 2}$ | $\frac{x^2 + 2x + 1}{x^2}$ | $\frac{x + 3}{x - 1}$ |
| $\frac{x^2 - 9}{x^2 - 4}$ | $\frac{x}{x^2 + 3x + 2}$ | $\frac{1}{x + 2}$ **Finish** |

**2.** Goal: Product of $3(x - 4)$

| | | |
|---|---|---|
| $\frac{18x^2 - 9x}{3x + 2}$ | $\frac{3x^2 + 20x + 12}{x^3 - 4x^2}$ | $\frac{27x^2}{6x}$ |
| $\frac{3x^2 - 10x - 8}{6x^2 + 33x - 18}$ **Start** | $\frac{6x^2 - 24x}{18x - 9}$ | $\frac{6x - 3}{6x + 4}$ **Finish** |
| $\frac{2x^2 - 11x - 6}{3x + 2}$ | $\frac{x^2 - 8x + 16}{4x^2 - 4x + 1}$ | $\frac{6x^2 + 24}{3x - 2}$ |

**For 3 and 4, draw a path from Start to Finish, dividing by each new expression along the way, to find a quotient equal to the goal. Use only vertical or horizontal moves, and the path cannot cross Itself.**

**3.** Goal: Quotient of $\frac{15}{x}$

| | | |
|---|---|---|
| $\frac{x^2}{2y}$ | $\frac{1}{15y}$ | $6y^2$ **Finish** |
| $\frac{y}{10}$ | $\frac{x}{5y}$ | $\frac{5}{6}$ |
| $\frac{2}{3x}$ **Start** | $\frac{xy}{3}$ | $\frac{2}{x^2y}$ |

**4.** Goal: Quotient of 1

| | | |
|---|---|---|
| $\frac{x - 3}{3x^2 - 10x + 3}$ | $\frac{12x^2 + 11x - 5}{x - 3}$ **Start** | $\frac{x^2 + 7x - 30}{3x^2 + 5x - 2}$ |
| $\frac{x + 2}{4x^2 + 45x + 50}$ | $\frac{3x^2 + 5x - 2}{x^2 + 7x - 30}$ | $\frac{4x^2 - 7x - 15}{x^2 + 12x + 20}$ |
| $\frac{3x^2 + 29x - 10}{x + 10}$ | $\frac{4x^2 + 45x + 50}{x + 2}$ **Finish** | $\frac{x - 3}{12x^2 + 11x - 5}$ |

Name ______________________ Date __________ Class __________

LESSON 12-4

# Problem Solving

## *Multiplying and Dividing Rational Expressions*

**A bag contains 5 more red marbles than blue marbles. Janis is going to pick two marbles out of the bag without looking, and without replacing the first marble. Let *x* represent the number of blue marbles.**

1. Write and simplify an expression that represents the probability that Janis will pick two blue marbles.

_______________________

2. Is the probability that Janis picks a red marble and then a blue marble the same as the probability that she picks a blue and then a red? Explain.

_______________________

_______________________

_______________________

_______________________

_______________________

3. Write and simplify an expression that represents the probability that Janis will pick two red marbles. Then find the probability of picking two red marbles if the bag contains 20 blue marbles.

_______________________

_______________________

**A square dart board has three regions, A, B, and C, as shown. Units are in inches. Select the best answer.**

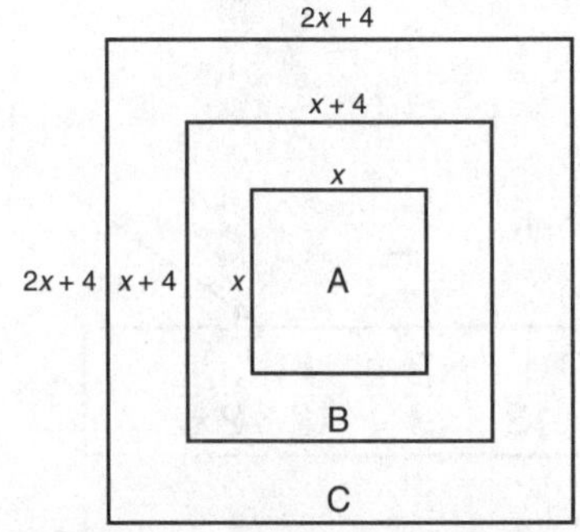


4. Which expression represents the probability of throwing two darts that land in region A and then region B?

   A $\frac{2x^2}{x+2}$

   B $\frac{x^2}{2(x+2)^3}$

   C $\frac{(x+4)^2}{16(x+2)^4}$

   D $\frac{x^2(x+4)^2}{4(x+2)^2}$

5. Divide two rational expressions to find how many times greater the probability of a dart landing in region C is than the probability of a dart landing in region A.

   F $\frac{3x+8}{x}$

   G $\frac{x}{2(x+1)}$

   H $\frac{4(x+2)^2}{x^2}$

   J $\frac{x(3x+8)}{8(x+2)}$

6. Which expression represents the probability of throwing two darts that both land in region B?

   A $\frac{4}{(x+2)^2}$

   B $\frac{x^2}{(x+4)^2}$

   C $\frac{(x+4)^4}{16(x+2)^4}$

   D $\frac{64(x+2)^2}{x^2(3x+8)^2}$

7. If $x = 2$ in., what is the probability of throwing two darts that both land in region A?

   F $\frac{1}{4}$

   G $\frac{1}{16}$

   H $\frac{1}{32}$

   J $\frac{1}{256}$

Name ______________________ Date __________ Class __________

**LESSON 12-4**

# Reading Strategies

## *Connecting Concepts*

The rules for multiplying and dividing rational expressions are the same as the rules for multiplying and dividing fractions. Look at the connections below.

**FRACTIONS** | **RATIONAL EXPRESSIONS**

**Rule 1:** $\frac{a}{b} \cdot \frac{c}{d} = \frac{ac}{bd}$

$\frac{8}{9} \cdot \frac{3}{5} = \frac{8 \cdot 3}{9 \cdot 5} = \ldots$

$\frac{m+2}{4} \cdot \frac{8}{3m+6} = \frac{8(m+2)}{4(3m+6)} = \ldots$

**Rule 2:** $\frac{a}{b} \cdot c = \frac{a}{b} \cdot \frac{c}{1}$ and $\frac{a}{b} \div c = \frac{a}{b} \div \frac{c}{1}$

$\frac{3}{4} \cdot 16 = \frac{3}{4} \cdot \frac{16}{1} = \ldots$

$\frac{6}{2x+10} \cdot (x^2 + 3x - 10) = \frac{6}{2x+10} \cdot \frac{x^2 + 3x - 10}{1} = \ldots$

**Rule 3:** $\frac{a}{b} \div \frac{c}{d} = \frac{a}{b} \cdot \frac{d}{c}$

$\frac{1}{12} \div \frac{2}{3} = \frac{1}{12} \cdot \frac{3}{2} = \ldots$

$\frac{3ab^4}{a^3} \div \frac{3ab}{ab^2} = \frac{3ab^4}{a^3} \cdot \frac{ab^2}{3ab} = \ldots$

**Finish multiplying and dividing the rational expressions given above so that each is simplified completely.**

**1.** Rule 1 ____________

**2.** Rule 2 ____________

**3.** Rule 3 ____________

**Multiply or divide. Simplify your answer.**

**4.** $\frac{3}{t+2} \cdot (t^2 - 4)$ ____________

**5.** $\frac{m+5}{4} \div (3m + 15)$ ____________

**6.** $\frac{x^2y^3}{8y} \cdot \frac{10x^4}{xy^2}$ ____________

Name ______________________________ Date ____________ Class ____________

LESSON 12-5

# Practice A

## *Adding and Subtracting Rational Expressions*

**Add or subtract. Simplify your answer.**

**1.** $\frac{6x}{5x^2} + \frac{4x}{5x^2}$

______________________

**2.** $\frac{t}{t^2 - 36} + \frac{6}{t^2 - 36}$

______________________

**3.** $\frac{c + 5}{3c + 3} - \frac{4}{3c + 3}$

______________________

**4.** $\frac{3b}{b^2 + 8b + 15} - \frac{b - 10}{b^2 + 8b + 15}$

______________________

**Find the LCM of the given expressions.**

**5.** $3x^2y^2$, $9x^3y$

______________________

**6.** $a^2 - 16$, $(a - 4)(a + 5)$

______________________

**Use a common denominator to add or subtract. Simplify your answer.**

**7.** $\frac{7}{3m} - \frac{2}{m}$

______________________

**8.** $\frac{x^2 + x}{x^2 + 6x + 5} + \frac{4}{x + 5}$

______________________

**9.** $\frac{3}{x - 10} + \frac{9}{10 - x}$

______________________

**10.** $\frac{3}{x} - \frac{4}{x^2 - 9x}$

______________________

**11.** Carlie biked 6 miles on mountain trails and 15 miles on the road. Her biking rate on the road was 3 times faster than that on the trails. Let *r* represent her rate on the trails in mi/h.

**a.** Write and simplify an expression that represents the number of hours she biked in all. $\left(\textit{Hint}: t = \frac{d}{r}\right)$ ______________________

**b.** Find the time if her rate on the trails was 3 mi/h. ______________________

Name ________________________ Date __________ Class __________

LESSON 12-5

# Practice B

## Adding and Subtracting Rational Expressions

**Add or subtract. Simplify your answer.**

1. $\frac{3m}{8m^3} + \frac{m}{8m^3}$

2. $\frac{x^2 - 6x}{x + 3} + \frac{4x - 15}{x + 3}$

3. $\frac{c^2 + c}{c^2 - 25} - \frac{c^2 + 5}{c^2 - 25}$

4. $\frac{6a - 1}{a^2 + 7a + 10} - \frac{2a - 9}{a^2 + 7a + 10}$

**Find the LCM of the given expressions.**

5. $4a^2b$, $6a$, $10b^3$

6. $x^2 + 5x + 6$, $(x + 3)(x - 1)$

**Add or subtract. Simplify your answer.**

7. $\frac{5}{3n} - \frac{2}{2n}$

8. $\frac{y^2 + 4y}{y^2 + 6y + 8} + \frac{3}{y + 2}$

9. $\frac{x + 2}{x^2 - 9} - \frac{1}{9 - x^2}$

10. $\frac{1}{6y^2 + 24y} - \frac{3}{y^2 - y - 20}$

11. Kendrick walked 1 mile, and then jogged 5 miles. His jogging speed was 4 times his walking speed $w$ in mi/h.

    a. Write and simplify an expression that represents Kendrick's total exercise time. ____________

    b. How many minutes did Kendrick exercise if his walking speed was 3 mi/h? ____________

Name ______________________ Date __________ Class __________

LESSON 12-5

# Practice C
## *Adding and Subtracting Rational Expressions*

**Add or subtract. Simplify your answer.**

1. $\frac{7y}{2y^3} + \frac{3y}{2y^3}$ ______________________

2. $\frac{8t + 5}{t^2 + t - 12} - \frac{5t - 7}{t^2 + t - 12}$ ______________________

3. $\frac{7m + 8}{4m^2 - 25} - \frac{m - 7}{4m^2 - 25}$ ______________________

4. $\frac{3}{p} + \frac{7}{4p}$ ______________________

5. $\frac{x^2 + 2x}{x^2 + 5x + 6} - \frac{5}{x + 3}$ ______________________

6. $\frac{4k}{k - 6} + \frac{k}{6 - k}$ ______________________

7. $\frac{x}{2x - 10} + \frac{1}{3x - 15}$ ______________________

8. $\frac{c}{c^2 - d^2} + \frac{d}{c^2}$ ______________________

9. $\frac{x}{x^2 - 4x - 5} + \frac{1}{x + 1} + \frac{1}{5}$ ______________________

10. $\frac{2}{t - 1} + \frac{t}{t^2 - 2t + 1} - \frac{3}{t + 1}$ ______________________

11. A boat sailed from a dock 35 miles up a river and back. The boat's average sailing rate is 6 times the speed of the river's current.

    a. Write and simplify an expression that represents the time it took the boat to make the round trip. ______________________

    b. The boat left the dock at 3:15 P.M. What time will the boat return if the rate of the river is 4.5 mi/h? ______________________

Name ________________________ Date ____________ Class ____________

LESSON 12-5

# Reteach

## Adding and Subtracting Rational Expressions

The rules for adding and subtracting rational expressions are the same as the rules for adding and subtracting fractions.

$$\frac{2}{7} + \frac{4}{7} = \frac{2+4}{7} = \frac{6}{7} \qquad \frac{4}{5} - \frac{1}{5} = \frac{4-1}{5} = \frac{3}{5}$$

**Add $\frac{5x+10}{x^2-16} + \frac{10}{x^2-16}$. Simplify your answer.**

$\frac{5x+10}{x^2-16} + \frac{10}{x^2-16}$

$\frac{(5x+10)+10}{x^2-16}$ *Add numerators.*

$\frac{5x+20}{x^2-16}$ *Add like terms.*

$\frac{5(x+4)}{(x+4)(x-4)}$ *Factor.*

$\frac{5\cancel{(x+4)}^1}{\cancel{(x+4)}^1(x-4)}$ *Simplify.*

$\frac{5}{x-4}$

**Subtract $\frac{2x}{2x^2+6} - \frac{x-3}{2x^2+6}$. Simplify your answer.**

$\frac{2x}{2x^2+6} - \frac{x-3}{2x^2+6}$

$\frac{2x-(x-3)}{2x^2+6}$ *Subtract numerators.*

$\frac{2x-x+3}{2x^2+6}$ *Distribute −1.*

$\frac{x+3}{2x^2+6}$ *Combine like terms.*

$\frac{x+3}{2(x+3)}$ *Factor.*

$\frac{\cancel{x+3}^1}{2\cancel{(x+3)}^1}$ *Simplify.*

$\frac{1}{2}$

**Add or subtract. Simplify your answer.**

1. $\frac{x+2}{x^2-100} + \frac{8}{x^2-100}$ ____________

2. $\frac{x^2-18}{x+6} + \frac{3x}{x+6}$ ____________

3. $\frac{x}{5x+30} + \frac{6}{5x+30}$ ____________

4. $\frac{x}{10x^2-20x} - \frac{2}{10x^2-20x}$ ____________

5. $\frac{x^2}{x^2+10x+25} - \frac{25}{x^2+10x+25}$ ____________

6. $\frac{x^2}{x+6} - \frac{12-4x}{x+6}$ ____________

Name ______________________ Date __________ Class __________

LESSON 12-5

# Reteach

## Adding and Subtracting Rational Expressions (continued)

Like fractions, you may need to multiply by a form of 1 to obtain common denominators.

$$\frac{1}{6}+\frac{4}{15}=\frac{1}{6}\left(\frac{5}{5}\right)+\frac{4}{15}\left(\frac{2}{2}\right)=\frac{5}{30}+\frac{8}{30}=\frac{13}{30}$$

**Add $\frac{3}{10x}+\frac{2}{5x^2}$.**
**Simplify your answer.**

$\frac{3}{10x}+\frac{2}{5x^2}$ *The LCD is $10x^2$.*

$\frac{3}{10x}\left(\frac{x}{x}\right)+\frac{2}{5x^2}\left(\frac{2}{2}\right)$ *Multiply each term by a form of 1.*

$\frac{3x}{10x^2}+\frac{4}{10x^2}$ *Multiply.*

$\frac{3x+4}{10x^2}$ *Add numerators.*

**Subtract $\frac{3}{x-4}-\frac{5}{4-x}$.**
**Simplify your answer.**

$\frac{3}{x-4}-\frac{5}{4-x}$ *The LCD is $x-4$.*

$\frac{3}{x-4}-\frac{5}{4-x}\left(\frac{-1}{-1}\right)$ *Multiply the second term by a form of 1.*

$\frac{3}{x-4}-\frac{-5}{x-4}$ *Multiply.*

$\frac{8}{x-4}$ *Add numerators.*

7. Identify the LCD of $\frac{5}{8x^2}$ and $\frac{13}{20x}$. ______________

8. Identify the LCD of $\frac{3}{5x}$ and $\frac{x}{x+7}$. ______________

**Add or subtract. Simplify your answer.**

9. $\frac{2x}{8x^3}+\frac{5}{12x}$

$\frac{2x}{8x^3}\left(\frac{\square}{\square}\right)+\frac{5}{12x}\left(\frac{\square}{\square}\right)$

______________

10. $\frac{5x+11}{x^2+5x+6}-\frac{4}{x+3}$

$\frac{5x+11}{(\square)(\square)}-\frac{4}{x+3}\left(\frac{\square}{\square}\right)$

______________

11. $\frac{4x}{6x}+\frac{2}{x+4}$

______________

12. $\frac{3x-6}{x^2-36}-\frac{2}{x+6}$

______________

13. $\frac{2x}{x^2+4x+3}+\frac{1}{x+1}$

______________

Name ______________________________ Date ____________ Class ____________

LESSON 12-5

# Challenge

## *Average Speed*

**At 8 A.M., Mr. Varhalla drives from home to work. Because of rush-hour traffic, he is only able to drive 40 mi/h.**

**At 3 P.M. he leaves work early to drive back home. Because he avoids rush-hour traffic, he is able to drive the same distance at 65 mi/h.**

1. Let $d$ represent the distance in miles between home and work. What is the total distance of Mr. Varhalla's round trip? ____________

2. Write an expression for the time in hours that it takes Mr. Varhalla to drive from home to work. ____________

3. Write an expression for the time in hours that it takes him to drive from work back home. ____________

4. Write (but don't simplify) an addition expression for the total time in hours of Mr. Varhalla's round trip. ____________

5. Because $d = rt$, you can find average speed for the entire round trip by dividing the total distance by the total time, or $r = \frac{d}{t}$. Write a complex fraction that represents Mr. Varhalla's average speed in miles per hour. ____________

6. Simplify the complex fraction to find Mr. Varhalla's average speed in miles per hour. Round to the nearest tenth. ____________

7. Is the average speed for the round trip the same as the average of the speeds 40 mi/h and 65 mi/h?

____________________________________________

**Use what you have learned in questions 1–7 to solve these problems. For each, first write a complex fraction that represents average speed and then simplify it to find the average speed. Round your answers to the nearest tenth.**

8. Teri drives 30 mi/h from home to the hardware store. After loading her truck with lumber, she drives 15 mi/h back home. What is Teri's average speed in miles per hour for the entire round trip?

____________

9. Kevin runs four laps around Piedmont Park at 10 mi/h. Then, to cool down, he walks one more lap at 3 mi/h. What is Kevin's average speed for the entire workout? (*Hint:* Let $d$ represent the distance of one lap.)

____________

Name ______________________ Date __________ Class __________

LESSON 12-5

# Problem Solving

## *Adding and Subtracting Rational Expressions*

**Adib is kayaking on the Peconic River in Long Island, New York. He paddles his kayak at an average rate of 3 mi/h, but does not know the rate of the river's current. Adib plans to kayak upstream 2 mi and then back downstream to his starting point.**

1. Let $x$ represent the rate of the current in the Peconic River in miles per hour. Write and simplify an expression for the total time of Adib's round trip.

______________________

2. If the rate of the river's current is 2 mi/h, how long will it take Adib to kayak round trip?

______________________

3. If the rate of the river's current is 3 mi/h, how long will it take Adib to kayak round trip? Explain what your answer could mean in this context.

______________________

______________________

**Select the best answer.**

4. Terry drives 2 mi on city streets, and 20 mi on the highway. Her speed on the highway is three times her speed on the city streets $r$, in miles per hour. Write and simplify and expression that represents the length of Terry's trip in hours.

   **A** $\frac{13}{3r}$ **C** $\frac{26}{3r}$

   **B** $\frac{22}{4r}$ **D** $\frac{62}{3r}$

5. Nahuel walks 1 km to school at a rate of $w$ km/h. When he gets to school, he realizes that he forgot his math book, so he jogs back home and back to school at a rate 4 km/h faster than his walking rate. Write and simplify an expression that represents the length of Nahuel's entire trip to school in hours.

   **F** $\frac{3}{2w}$ **H** $\frac{2(w+1)}{w(w+2)}$

   **G** $\frac{3}{2(w+2)}$ **J** $\frac{3w+4}{w(w+4)}$

6. A test consists of 4 free response questions and 50 multiple choice questions. The test's writers assume an average student can do $x$ free response questions per hour, and four times as many multiple choice questions per hour. If an average student can do 6 free response questions per hour, how long is the test?

   **A** $1\frac{4}{5}$ h **C** $3\frac{3}{8}$ h

   **B** $2\frac{3}{4}$ h **D** $5\frac{2}{3}$ h

7. In a carnival game, Beth runs 100 m to a table, picks up water balloons, and runs back to the starting line. Beth originally runs to the table at a rate of $y$ m/s, but runs 1 m/s slower when she carries the balloons. If Beth originally runs 5 m/s, how long, to the nearest second, does it take her to finish?

   **F** 10 s **H** 37 s

   **G** 22 s **J** 45 s

Name ____________________ Date ________ Class ________

LESSON 12-5

# Reading Strategies

## *Follow a Procedure*

When adding or subtracting rational expressions, there may be several steps that need to be taken. Look at the example below.

**Add** $\frac{3}{9x} + \frac{x}{6x^3}$.

| | |
|---|---|
| $9x = 3 \cdot 3 \cdot x$<br>$6x^3 = 2 \cdot 3 \cdot x \cdot x \cdot x$<br>LCD: $2 \cdot 3 \cdot 3 \cdot x \cdot x \cdot x = 18x^3$ | 1. Identify the least common denominator, LCD. |
| $\frac{3}{9x}\left(\frac{2x^2}{2x^2}\right) + \frac{x}{6x^3}\left(\frac{3}{3}\right)$ | 2. Multiply expressions by appropriate form of 1. |
| $\frac{6x^2}{18x^3} + \frac{3x}{18x^3}$ | 3. Write each expression with the new LCD. |
| $\frac{6x^2 + 3x}{18x^3}$ | 4. Add/subtract numerators. Keep denominator. |
| $\frac{3x(2x + 1)}{18x^3}$ | 5. Factor the numerator and/or denominator. |
| $\frac{2x + 1}{6x^2}$ | 6. Simplify as needed. |

**Answer each question about the procedure shown above.**

1. If the two expressions have the same denominator, which step can you start with? ____________

2. Why was $\left(\frac{2x^2}{2x^2}\right)$ the "appropriate form of 1" for the first expression?

_______________________________________________

3. How would the answer to this problem change if it were subtraction instead of addition?

_______________________________________________

**Add or subtract.**

4. $\frac{x + 5}{x^2 - 9} - \frac{2}{x^2 - 9}$

5. $\frac{1}{4x} + \frac{4x}{6x^2}$

____________________ ____________________

Name ______________________ Date __________ Class __________

LESSON 12-6

# Practice A
## *Dividing Polynomials*

**Divide each term of the polynomial by the monomial.**

1. $(12m^3 + 4m^2) \div 4m$

   $\frac{12m^3}{\square} + \frac{4m^2}{\square}$

   ______________________

2. $(40y^5 + 24y^3 - 20y^2) \div 4y^2$

   ______________________

3. $(15x^5 + 15x^2 - 10x) \div 5x^2$

   ______________________

4. $(6w^7 - 2w^3 - 7) \div 2w^4$

   ______________________

**Divide each polynomial by the binomial.**

5. $(k^2 + 14k + 45) \div (k + 9)$

   ______________________

6. $(x^2 - 5x - 14) \div (x + 2)$

   ______________________

**Divide using long division. There will not be a remainder.**

7. $a + 3\overline{)a^2 + 4a + 3}$

   ______________________

8. $t + 4\overline{)2t^2 + 11t + 12}$

   ______________________

**Divide using long division. There will be a remainder.**

9. $(c^2 + 7c + 14) \div (c + 4)$

   ______________________

10. $(k^2 + 5) \div (k - 2)$

    (*Hint*: Rewrite $k^2 + 5$ as $k^2 + 0k + 5$.)

    ______________________

Name ______________________ Date __________ Class __________

LESSON 12-6

# Practice B
## *Dividing Polynomials*

**Divide.**

**1.** $(15c^3 + 3c^2) \div 3c$

______________________

**2.** $(20b^4 - 12b + 4) \div 4b$

______________________

**3.** $(27q^6 - 3q^3 + 18) \div 9q^5$

______________________

**4.** $(15t^4 - 30t^2 + 6) \div 15t^3$

______________________

**5.** $(d^2 - 4d - 77) \div (d - 11)$

______________________

**6.** $(x^2 - 12x + 27) \div (x - 3)$

______________________

**7** $(9p^2 + 6p + 1) \div (3p + 1)$

______________________

**8.** $(4b^2 + b - 3) \div (b + 1)$

______________________

**Divide using long division.**

**9.** $\dfrac{m^2 + 4m - 12}{m + 6}$

______________________

**10.** $(12y^2 + 31y + 14) \div (y + 2)$

______________________

**11.** $(t^2 + t - 6) \div (t - 1)$

______________________

**12.** $(3p^3 + 4p - 6) \div (p + 2)$

______________________

Name ______________________ Date __________ Class __________

LESSON 12-6

# Practice C
## *Dividing Polynomials*

**Divide.**

**1.** $(3x^2 + x) \div 3x$

______________________

**2.** $(12x^2 - 7x + 16) \div -2x^2$

______________________

**3.** $(k^6 - 2k^5 + 10k^3 - 20) \div 5k^5$

______________________

**4.** $(-3m^5n^4 + 12m^3n^2 - mn) \div -6m^2n^2$

______________________

**5.** $(4t^2 + 17t - 42) \div (t + 6)$

______________________

**6.** $(24x^2 + x - 10) \div (8x - 5)$

______________________

**7.** $(48b^2 + 32b + 5) \div (4b + 1)$

______________________

**8.** $(3y^3 + 15y^2 + y + 5) \div (y + 5)$

______________________

**Divide using long division.**

**9.** $\dfrac{2n^2 - 11n + 12}{2n - 3}$

______________________

**10.** $(3a^2 + 7a - 20) \div (a + 4)$

______________________

**11.** $(2y^2 + 3y - 5) \div (y - 3)$

______________________

**12.** $(4t^2 + 3t^3 - 8) \div (6 + t)$

______________________

Name ______________________ Date __________ Class __________

LESSON 12-6

# Reteach

## Dividing Polynomials

To divide a polynomial by a monomial, first write the division as a rational expression.

**Divide $(12x^2 + 9x) \div 3x$.**

$\frac{12x^2 + 9x}{3x}$ *Rewrite as a rational expression.*

$\frac{12x^2}{3x} + \frac{9x}{3x}$ *Divide each term by 3x.*

$\frac{\cancel{12}^4 \cancel{x^2}^1}{_1\cancel{3} \cancel{x}_1} + \frac{\cancel{9}^3 \cancel{x}^1}{_1\cancel{3} \cancel{x}_1}$ *Divide out common factors.*

$4x + 3$ *Simplify.*

To divide a polynomial by a binomial, try to factor and divide out common factors.

**Divide $\frac{x^2 + 6x + 5}{x + 1}$.**

$\frac{x^2 + 6x + 5}{x + 1}$

$\frac{(x + 1)(x + 5)}{x + 1}$ *Factor the numerator.*

$\frac{\cancel{(x + 1)}^1 (x + 5)}{\cancel{x + 1}_1}$ *Divide out common factors.*

$x + 5$ *Simplify.*

You may find that a rational expression does not divide evenly.

**Divide $(5x^2 + 10x + 3) \div 5x$.**

$\frac{5x^2 + 10x + 3}{5x}$ *Rewrite as a rational expression.*

$\frac{5x^2}{5x} + \frac{10x}{5x} + \frac{3}{5x}$ *Divide each term by 5x.*

$x + 2 + \frac{3}{5x}$

**Divide.**

1. $(14x^3 + 6x) \div 2x$

2. $(6x^4 + 12x^2 + 18x) \div 6x^2$

3. $(10x^3 + 12x^2 + 2x) \div 2x$

____________ ____________ ____________

4. $(x^2 - 49) \div (x + 7)$

5. $(3x^2 - 15x) \div (x - 5)$

6. $(x^2 - 5x - 24) \div (x - 8)$

____________ ____________ ____________

Name ______________________ Date ____________ Class ____________

LESSON 12-6

# Reteach

## Dividing Polynomials (continued)

You can use long division to divide a polynomial by a binomial.

**Divide $(2x^2 + 2x + 7) \div (x + 3)$.**

$x + 3\overline{)2x^2 + 2x + 7}$ *Write in long division form.*

$$\begin{array}{r} 2x - 4 \\ x + 3\overline{)2x^2 + 2x + 7} \\ -(2x^2 + 6x) \\ \hline -4x + 7 \\ -(-4x - 12) \\ \hline 19 \end{array}$$

*Think: $x \cdot ? = 2x^2$. Use $2x$.*

*Think: $x \cdot ? = -4x$. Use $-4$.*

$(2x^2 + 2x + 7) \div (x + 3) = 2x - 4 + \frac{19}{x + 3}$

Use a zero coefficient if the polynomial is missing a term.

**Divide $(x^2 - 4) \div (x + 2)$.**

$x + 2\overline{)x^2 + 0x - 4}$ *Write in long division form.*

$$\begin{array}{r} x - 2 \\ x + 2\overline{)x^2 + 0x - 4} \\ -(x^2 + 2x) \\ \hline -2x - 4 \\ -(-2x - 4) \\ \hline 0 \end{array}$$

*Think: $x \cdot ? = x^2$. Use $x$.*

*Think: $x \cdot ? = -2x$. Use $-2$.*

$(x^2 - 4) \div (x + 2) = (x - 2)$

**Divide using long division.**

7. $(x^2 - 4x - 12) \div (x + 2)$

$$\begin{array}{r} \square - \square \\ x + 2\overline{)x^2 + 4x - 12} \\ -(\square) \\ \hline \square - \square \\ -(\square) \\ \hline 0 \end{array}$$

________________________

8. $(x^2 + 6x + 3) \div (x + 4)$

$$\begin{array}{r} \square + \square \\ x + 4\overline{)x^2 + 6x + 3} \\ -(\square) \\ \hline \square + \square \\ -(\square) \\ \hline -5 \end{array}$$

________________________

9. $(x^2 - 25) \div (x + 5)$

________________________

10. $(x^2 + 5x + 4) \div (x + 2)$

________________________

Name ______________________ Date __________ Class __________

LESSON 12-6

# Challenge

## *The Remainder Theorem*

**For 1–5, use the polynomial $P(x) = 2x^2 + 3x - 5$.**

1. Use long division to divide $P(x) \div (x - 2)$.

   a. What value do you get for the remainder? __________

   b. Now evaluate $P(2)$. What value do you get? __________

2. Use long division to divide $P(x) \div (x - 1)$.

   a. What value do you get for the remainder? __________

   b. Now evaluate $P(1)$. What value do you get? __________

3. Use long division to divide $P(x) \div x$.

   a. What value do you get for the remainder? __________

   b. Now evaluate $P(0)$. What value do you get? __________

4. Use long division to divide $P(x) \div (x + 1)$.

   a. What value do you get for the remainder? __________

   b. Now evaluate $P(-1)$. What value do you get? __________

5. Use long division to divide $P(x) \div (x + 2)$.

   a. What value do you get for the remainder? __________

   b. Now evaluate $P(-2)$. What value do you get? __________

6. Look for a pattern in your results and complete this theorem.

   **Polynomial Remainder Theorem:** If a polynomial $P(x)$ is divided by $(x - r)$, then the remainder is a constant given by __________.

7. Your results in question 2 are unique. What do they tell you about the binomial $(x - 1)$?

   ________________________________________

8. Given $Q(x) = 6x^4 - 31x^3 - 49x^2 + 104x + 60$, explain how you could determine whether a binomial of the form $(x - r)$ is a factor of $Q(x)$ without doing long division.

   ________________________________________

9. **a.** Is $(x - 1)$ a factor of $Q(x)$? Justify.

   ________________________________________

   **b.** Is $(x + 2)$ a factor of $Q(x)$? Justify.

   ________________________________________

Name ______________________ Date __________ Class __________

LESSON 12-6

# Problem Solving

## Dividing Polynomials

**Write the correct answer.**

**1.** The area of a rectangle is $x^2 + 2x - 8\text{m}^2$, and the width is $x - 2$ m. Find the length.

________________________

**2.** On the dartboard below, the area of the outside ring is the difference between the area of the two largest circles, or $\pi(x + 4)^2 - \pi(x + 2)^2$ in$^2$. The area of the bull's-eye is $\pi x^2$in$^2$. Expand and divide $[\pi(x + 4)^2 - \pi(x + 2)^2] \div \pi x^2$ to find the ratio of the area of the outside ring to the bull's-eye.

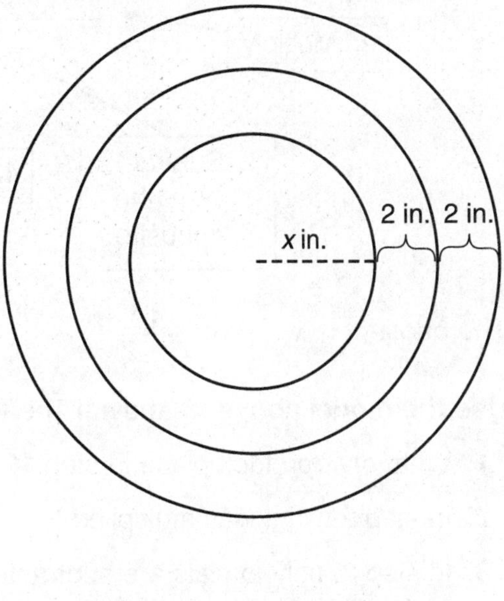


________________________

**3.** The area of a rectangular piece of paper is $4n^2 - 4n - 9$ cm$^2$. Then a triangle with area $2n + 6$ cm$^2$ is cut from one corner of the paper. What is the ratio of the area of the paper that remains to the area of the triangle that was removed? (*Hint:* First subtract to find the area of the paper that remains.)

________________________

**Select the best answer.**

**4.** The area of a triangle is $8a^2 - 10a$ ft$^2$, and the height is $4a$ ft. Find the base. (*Hint:* Solve $A = \frac{1}{2}bh$ for $b$.)

**A** $a - \frac{5a}{4}$ ft  **C** $4a - \frac{5a}{2}$ ft

**B** $2a - \frac{5a}{2}$ ft  **D** $4a - 5$ ft

**5.** The volume of a rectangular prism is $2x^3 + 9x^2 - 11x - 30$ cm$^3$, and the height is $x + 5$ cm. Find the area of the base. ($V = BH$, where $B$ is the area of the base and $H$ is the height.)

**F** $2x^2 - x - 6$ cm$^2$

**G** $2x^2 - x - 16 + \frac{-110}{x + 5}$ cm$^2$

**H** $2x^2 + 4x - 16 + \frac{-35}{x + 5}$ cm$^2$

**J** $2x^2 + 19x + 84 + \frac{390}{x + 5}$ cm$^2$

**6.** The volume of a cylinder is $V = \pi r^2 h$, where $r$ is the radius and $h$ is the height. The volume of a certain cylinder is $\pi(n^3 - 6n^2 - 36n + 216)$ in$^3$, and the radius is $n - 6$ in. Find the height. (*Hint:* To divide by $r^2$, divide by $r$ twice.)

**A** $n$ in.  **C** $n + 6$ in.

**B** $n - 6$ in.  **D** $n^2 - 36$ in.

Name ______________________ Date __________ Class __________

LESSON 12-6

# Reading Strategies
## *Use a Model*

The process for dividing polynomials is the same as the process for dividing whole numbers. The steps are shown in the model below.

**Divide $(x^2 + 5x - 4) \div (x + 2)$.**

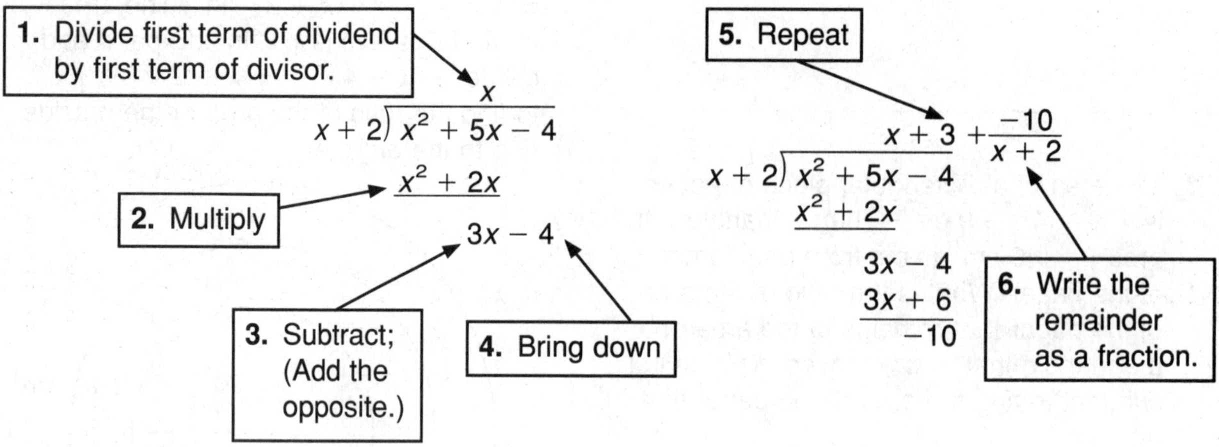


The quotient is $x + 3 + \frac{-10}{x + 2}$.

**Use the model above to answer the following.**

1. What division took place in step 1? ______________
2. In step 2, what was multiplied? ______________
3. In step 3, polynomials are subtracted by "adding the opposite."
   What is the opposite of $x^2 + 2x$? ______________
4. What term was brought down in step 4? ______________
5. In step 5, what is meant by "repeat"? ______________

______________________________

**Divide using long division.**

6. $(x^2 + 8x + 15) \div (x + 3)$

7. $(x^2 - 2x + 7) \div (x + 3)$

______________ ______________

Name ______________________ Date __________ Class __________

LESSON 12-7

# Practice A

## *Solving Rational Equations*

**Solve using cross products. Check your answer.**

1. $\frac{6}{b+4} = \frac{4}{b}$ ____________

2. $\frac{2}{x} = \frac{3}{x-5}$ ____________

**Solve using the LCD. Check your answer.**

3. $\frac{9}{c} - \frac{8}{c} = \frac{1}{4}$ ____________

4. $\frac{2}{n+1} + \frac{3}{n} = \frac{9}{2n}$ ____________

5. $\frac{9}{2x} + \frac{7}{x} = \frac{12x-1}{2x}$ ____________

6. $\frac{3}{x-2} + \frac{3}{x+2} = \frac{9}{x+2}$ ____________

**Solve using cross products. Identify any extraneous solutions.**

7. $\frac{2}{x-4} = \frac{x-1}{x-4}$ ____________

8. $\frac{x-2}{x+2} = \frac{3}{x}$ ____________

9. Marlena can rake the leaves in 4 hours. Joshua can rake the same amount of leaves in 2 hours. How long will it take them to rake the leaves if they work together? ____________

**One number is 4 times another number. What are the numbers if the sum of their reciprocals is $\frac{1}{4}$?**

10. If one of the numbers is $x$, what is the other number? ____________

11. Write the reciprocal of each number. ____________

12. Write and solve an equation showing that the sum of the reciprocals is $\frac{1}{4}$. ____________

____________

13. What are the two numbers? ____________

Name ____________________ Date __________ Class __________

LESSON 12-7

# Practice B

## *Solving Rational Equations*

**Solve. Check your answer.**

1. $\frac{6}{t+3} = \frac{4}{t}$

2. $\frac{3}{m} = \frac{4}{m-2}$

3. $\frac{a}{4} + \frac{1}{2} = \frac{2}{3}$

4. $\frac{3}{2x} - \frac{3}{x-2} = \frac{1}{2x}$

5. $\frac{3}{2x} + \frac{5}{x} = \frac{13}{x+4}$

6. $\frac{3}{x} + \frac{3x+1}{x^2} = \frac{13}{x^2}$

**Solve. Identify any extraneous solutions.**

7. $\frac{8}{x-2} = \frac{x+3}{x-2}$

8. $\frac{-2}{x-1} = \frac{x-8}{x+1}$

9. Caroline can paint a fence in 6 hours. Her sister Lily can paint the same fence in 4 hours. How long will it take them to paint the fence if they work together?

10. Jalon bicycled against the wind for 10 miles in the same time he bicycled with the wind for 25 miles. The wind speed was 4 mi/h. What was Jalon's average bicycling speed? (*Hint*: Use $t = \frac{d}{r}$.)

11. There are two positive numbers. The second number is 6 less than the first number. When the reciprocal of the second number is subtracted from the reciprocal of the first, the difference is $-\frac{3}{8}$. Find the first number.

Name ______________________ Date __________ Class __________

LESSON 12-7

# Practice C

## *Solving Rational Equations*

**Solve. Check your answer.**

**1.** $\frac{4}{y} = \frac{1}{y + 6}$ ______________________

**2.** $\frac{5}{3x + 1} = 2$ ______________________

**3.** $\frac{4}{c + 2} + \frac{5}{c} = \frac{21}{3c}$ ______________________

**4.** $\frac{2}{n} = \frac{15}{n^2} - 1$ ______________________

**5.** $\frac{5}{6} - \frac{1}{x} = \frac{3x + 8}{6x}$ ______________________

**6.** $\frac{4}{x + 3} + \frac{3}{x} = \frac{4x + 1}{5x}$ ______________________

**7.** $\frac{3}{x + 3} = \frac{x - 2}{x + 3}$ ______________________

**8.** $\frac{x}{x^2 - 8} = \frac{2}{x}$ ______________________

**9.** Trey can clean a room in 40 minutes. His brother Marcus can clean the same room in 20 minutes. How long will it take them to clean the room if they work together? ______________________

**10.** A boat traveled against the current for 70 miles for the same time it traveled with the current for 95 miles. What was the boat's average speed if the rate of the current was 5 mi/h? ______________________

**11.** The reciprocal of a number minus the reciprocal of two less than the number is $-\frac{2}{15}$. Find the number. ______________________

**12.** Pump A can fill a tank 3 hours faster than Pump B. If both pumps work together, it takes 8 hours to fill the tank. Working alone, how long would it take Pump A to fill the tank? Round to the nearest tenth of an hour. (*Hints*: Use the Quadratic Formula and make sure times are positive for both pumps.) ______________________

Name ______________________ Date __________ Class __________

LESSON 12-7

# Reteach

## *Solving Rational Equations*

A **rational equation** is an equation that contains one or more rational expressions. Some rational equations are proportions and can be solved using cross products. Solutions to all rational equations must be checked.

**Solve $\frac{4}{x-3} = \frac{2}{x}$.**

$\frac{4}{x-3} \quad \frac{2}{x}$ (cross multiply)

$4(x) = 2(x-3)$ *Multiply.*

$4x = 2x - 6$ *Distribute.*

$-2x \quad -2x$ *Add $-2x$ to both sides.*

$2x = -6$

$\frac{2x}{2} = \frac{-6}{2}$ *Divide.*

$x = -3$ *Simplify.*

***Check:***

| $\frac{4}{x-3}$ | $= \frac{2}{x}$ |
|---|---|
| $\frac{4}{(-3)-3}$ | $\frac{2}{(-3)}$ |
| $\frac{4}{-6}$ | $\frac{2}{-3}$ |
| $-\frac{2}{3}$ | $-\frac{2}{3}$ ✓ |

The solution is $-3$.

**Solve $\frac{x-4}{x^2-4} = \frac{-2}{x-2}$.**

$\frac{x-4}{x^2-4} \quad \frac{-2}{x-2}$ (cross multiply)

$(x-4)(x-2) = -2(x^2-4)$

$x^2 - 6x + 8 = -2x^2 + 8$

$+2x^2 \quad +2x^2$

$3x^2 - 6x + 8 = 8$

$-8 \quad -8$

$3x^2 - 6x = 0$

$3x(x-2) = 0$ *Zero Product*

$x = 0$ or $x = 2$ *Property*

***Check:***

| $\frac{x-4}{x^2-4}$ | $= \frac{-2}{x-2}$ | $\frac{x-4}{x^2-4}$ | $= \frac{-2}{x-2}$ |
|---|---|---|---|
| $\frac{(0)-4}{(0)^2-4}$ | $\frac{-2}{(0)-2}$ | $\frac{(2)-4}{(2)^2-4}$ | $\frac{-2}{(2)-2}$ |
| $\frac{-4}{-4}$ | $\frac{-2}{-2}$ | $\frac{-2}{0}$ | $\frac{-2}{0}$ ✗ |
| 1 | 1 ✓ | undefined | |

The only solution is 0.

**Solve. Check your answer.**

1. $\frac{3}{x+2} = \frac{4}{x+1}$ ______________

2. $\frac{x}{6} = \frac{x}{x+4}$ ______________

3. $\frac{5}{x+3} = \frac{6}{x+1}$ ______________

4. $\frac{2x}{6} = \frac{x}{x+1}$ ______________

5. $\frac{8}{x^2-64} = \frac{1}{x-8}$ ______________

6. $\frac{x+2}{x-2} = \frac{4}{x-4}$ ______________

Name ______________________ Date ___________ Class ___________

LESSON 12-7

# Reteach

## Solving Rational Equations (continued)

If a rational equation contains sums or differences of rational expressions, multiply every term by the LCD to eliminate the fractions.

**Solve $\frac{1}{x} + \frac{2}{x+3} = \frac{3}{2x}$.**

$\frac{1}{x} + \frac{2}{x+3} = \frac{3}{2x}$ — *The LCD is 2x(x + 3).*

$2x(x+3)\left(\frac{1}{x}\right) + 2x(x+3)\left(\frac{2}{x+3}\right) = 2x(x+3)\left(\frac{3}{2x}\right)$ — *Multiply every term by the LCD.*

$2\cancel{x}^{1}(x+3)\left(\frac{1}{\cancel{x}^{1}}\right) + 2x\cancel{(x+3)}^{1}\left(\frac{2}{\cancel{x+3}^{1}}\right) = \cancel{2x}^{1}(x+3)\left(\frac{3}{\cancel{2x}^{1}}\right)$ — *Divide out common factors.*

$2(x+3) + 2x(2) = (x+3)(3)$ — All fractions have been eliminated.

$2x + 6 + 4x = 3x + 9$

$6x + 6 = 3x + 9$

$\underline{-3x} \quad \underline{-3x}$

$3x + 6 = 9$

$\underline{-6} \quad \underline{-6}$

$\frac{3x}{3} = \frac{3}{3}$

$x = 1$

***Check:***

$\frac{1}{x} + \frac{2}{x+3} = \frac{3}{2x}$

$\frac{1}{1} + \frac{2}{1+3} = \frac{3}{2(1)}$

$1 + \frac{1}{2} = \frac{3}{2}$

$\frac{3}{2} = \frac{3}{2}$ ✓

**Solve. Check your answer.**

**7.** $\frac{3}{x} + \frac{2}{x+2} = \frac{4}{x}$

LCD: ____________________

____________________

**8.** $\frac{10}{x} - \frac{8}{x} = \frac{6}{x+1}$

LCD: ____________________

____________________

**9.** $\frac{1}{4} + \frac{1}{4x} = \frac{5}{x^2}$

LCD: ____________________

____________________

**10.** $\frac{22}{(x+1)(x-1)} + \frac{2}{x+1} = \frac{8}{x-1}$

LCD: ____________________

____________________

Name ______________________ Date ____________ Class ____________

LESSON 12-7

# Challenge

## Exploring Reciprocals, Rational Equations, and Integer Solutions

In Lesson 12.7, you learned how to solve rational equations in one variable. Each equation below is a rational equation in two variables, $x$ and $y$.

$$\frac{1}{x} + \frac{1}{y} = 1 \qquad \frac{1}{x} + \frac{1}{y} = 2 \qquad \frac{1}{x} + \frac{1}{y} = \frac{1}{2}$$

In the exercises on this page, assume that $x$ and $y$ are positive integers.

**Find one solution to each equation of the form $\frac{1}{x} + \frac{1}{y} = \frac{1}{n}$, where $n$ is a positive integer.**

1. $\frac{1}{x} + \frac{1}{y} = 1$ ____________
2. $\frac{1}{x} + \frac{1}{y} = \frac{1}{2}$ ____________
3. $\frac{1}{x} + \frac{1}{y} = \frac{1}{4}$ ____________
4. $\frac{1}{x} + \frac{1}{y} = \frac{1}{6}$ ____________
5. $\frac{1}{x} + \frac{1}{y} = \frac{1}{8}$ ____________
6. $\frac{1}{x} + \frac{1}{y} = \frac{1}{100}$ ____________

7. Suppose that you are given $\frac{1}{x} + \frac{1}{y} = \frac{1}{n}$, where $n$ is an even positive integer. Based on your work in Exercises 1–6, write formulas in terms of $n$ for $x$ and for $y$. ____________

8. **a.** Let $\frac{1}{x} + \frac{1}{y} = 2$. Show that $x = 1$ and $y = 1$ are solutions to the equation by using substitution. ____________

   **b.** If $x$ and $y$ are positive integers other than 1, then $0 < \frac{1}{x} < 1$ and $0 < \frac{1}{y} < 1$. Explain how this fact helps you show that there are no positive integer solutions to $\frac{1}{x} + \frac{1}{y} = 2$ other than $x = 1$ and $y = 1$.

   ______________________________________________

   ______________________________________________

   **c.** How does the reasoning in Part b help you show that, when $n$ is an integer greater than 2 and $x$ and $y$ are positive integers, the equation $\frac{1}{x} + \frac{1}{y} = n$ has no integer solutions?

   ______________________________________________

**Use trial and error to help you decide whether the given equation has any integer solutions. If so, state one solution.**

9. $\frac{1}{x} + \frac{1}{y} = \frac{1}{3}$ ____________
10. $\frac{1}{x} + \frac{1}{y} = \frac{1}{5}$ ____________

Name ______________________ Date __________ Class __________

LESSON 12-7

# Problem Solving

## *Solving Rational Equations*

**Write the correct answer.**

**1.** Alex can remodel a bathroom in 2 days. Abraham can remodel the same bathroom in 4 days. How long will it take them to remodel the bathroom if they work together?

______________________

**2.** When only the cold faucet is fully turned on, a bathtub fills in 8 minutes. When only the hot faucet is fully turned on, the same bathtub fills in 12 minutes. If both the cold and hot faucets are fully turned on, how long will it take to fill the bathtub?

______________________

**3.** 100 mL of a solution contains 30 mL of acid; this is called a 30% acid solution because $\frac{\text{part}}{\text{whole}} = \frac{30}{100} = 0.30$. How many milliliters of acid x would need to be added to the solution to turn it into a 50% acid solution? (*Hint*: Adding acid increases both the volume of acid and the volume of the complete solution.)

______________________

**4.** Markus jogs 4 mi around a track at an average rate of *r* mi/h. Then he cools down by walking 1 mi at a rate 3 mi/h slower. His whole workout lasts 1 h. At what rates did Markus jog and walk?

______________________

**Select the best answer.**

**5.** Ms. Spinoni can prepare a mass-mailing of 500 letters in 10 hours. Mr. Harris can prepare a mass-mailing of 500 letters in 15 hours. How long will it take them to prepare a mass-mailing of 1000 letters if they work together? (*Hint:* The complete job is twice as big.)

**A** 6 h

**B** 12 h

**C** $12\frac{1}{2}$ h

**D** 25 h

**6.** At a cafeteria, the automatic ice machine can completely fill itself in 20 minutes. During lunch time, customers can completely empty the ice machine in 30 minutes. At the start of lunch time, the ice machine is completely empty and it starts making ice at the same time that customers start taking ice. How long will it take for the machine to be completely full? (*Hint:* The customers *take away* ice.)

**F** 12 min

**G** 20 min

**H** 30 min

**J** 60 min

**7.** To attend a family reunion, Beth drives 100 mi from Fresno and Clara drives 220 mi from San Jose. Both women drive the same speed in miles per hour, but Clara's drive takes 2 h longer than Beth's. Find the length of Beth's drive.

**A** 1 h

**B** $1\frac{2}{3}$ h

**C** $2\frac{1}{5}$ h

**D** $3\frac{2}{3}$ h

Name ______________________ Date __________ Class __________

LESSON 12-7

# Reading Strategies

## *Use a Concept Map*

**Rational Equation**

**Definition**

A *rational equation* is an equation that has one or more rational expressions. That is, the expressions have at least one variable in the denominator.

**Methods of Solving**

- Cross-products – when the equation forms a proportion
- LCD – multiply each term by the LCD to eliminate the fractions

**Examples**

$$\frac{8}{x} = \frac{4}{x - 1}$$

$$\frac{1}{m} + \frac{3}{m + 2} = \frac{2}{m - 4}$$

**Extraneous Solutions**

Rational equations can have *extraneous solutions.* These are not solutions to the original equation. Solutions that are excluded values (values that make the denominator zero) are extraneous. Always check your answers.

**Answer the following.**

1. Which example above can be solved using the cross-products? __________

2. Tell whether or not each of the following is a rational equation.

a. $\frac{5}{6} = \frac{x - 2}{3}$ __________ b. $\frac{4}{x} = \frac{2}{x^2 - 9}$ __________ c. $\frac{x}{5} + \frac{1}{x^2} = \frac{5}{4}$ __________

3. What is the purpose of multiplying a rational equation by the LCD?

__________________________________________

4. Liam solved the rational equation $\frac{x + 4}{x - 3} = \frac{3x}{x - 3}$ and got solutions 2 and 3. Is either solution extraneous? Explain. __________________________________________

__________________________________________

**Solve each rational equation. Check your answer.**

5. $\frac{3}{m + 2} = \frac{1}{m - 4}$

6. $\frac{4}{c - 4} = \frac{c}{4} + \frac{c}{c - 4}$

__________________ __________________

LESSON 12-1

## Practice A

### Inverse Variation

**Tell whether each relationship is an inverse relation. Explain.**

1.

| x | y |
|---|---|
| 1 | 30 |
| 2 | 15 |
| 3 | 10 |

$xy =$ 30
$xy =$ 30
$xy =$ 30

yes; $xy$ is constant.

2.

| x | y |
|---|---|
| 9 | 2 |
| 7 | 3 |
| 5 | 4 |

$xy =$ 18
$xy =$ 21
$xy =$ 20

no; $xy$ is not constant.

3. $x + y = 4$

no; the equation cannot be written in the form $y = \frac{k}{x}$.

4. $xy = 7$

yes; the equation can be written in the form $y = \frac{k}{x}$.

5. Write and graph the inverse variation in which $y = 6$ when $x = 2$ by following the steps below.

a. Find the constant of variation.

$k =$ 12

b. Write the inverse variation equation.

$y = \frac{12}{x}$

c. Make a table of values and graph.

| x | −6 | −4 | −2 | 0 | 2 | 4 | 6 |
|---|---|---|---|---|---|---|---|
| y | −2 | −3 | −6 | und. | 6 | 3 | 2 |

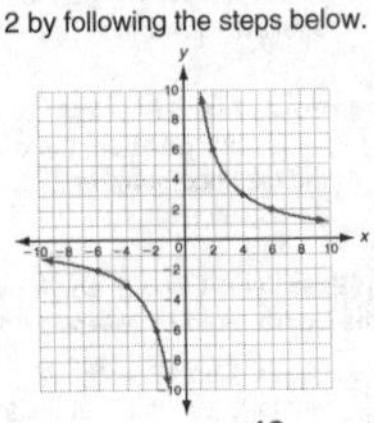

6. Let $x_1 = 5$, $y_1 = 4$, and $x_2 = 2$. Let $y$ vary inversely as $x$. Find $y_2$. 10

7. Let $x_1 = 3$, $y_1 = 7$, and $y_2 = 21$. Let $y$ vary inversely as $x$. Find $x_2$. 1

8. While traveling in a car, the speed of the car $x$ is inversely proportional to the time $y$ it takes to travel a certain distance. At 20 mi/h, it takes 10 minutes to travel to work. How many minutes would it take traveling 25 mi/h?

a. List the given information.

$x_1 =$ 20, $y_1 =$ 10, $x_2 =$ 25

b. Set up and solve an equation to find $y_2$.

$(20)(10) = 25(y_2)$

$y_2 =$ 8

It will take 8 minutes.




LESSON 12-1

## Practice B

### Inverse Variation

**Tell whether each relationship is an inverse relation. Explain.**

1.

| x | y |
|---|---|
| 2 | 12 |
| 3 | 8 |
| 4 | 6 |

yes; $xy$ is constant.

2.

| x | y |
|---|---|
| 1 | 4 |
| 2 | 8 |
| 3 | 12 |

no; $xy$ is not constant.

3. $x = \frac{y}{5}$

no; the equation cannot be written in the form $y = \frac{k}{x}$.

4. $xy = 8$

yes; the equation can be written in the form $y = \frac{k}{x}$.

5. Write and graph the inverse variation in which $y = 3$ when $x = 2$.

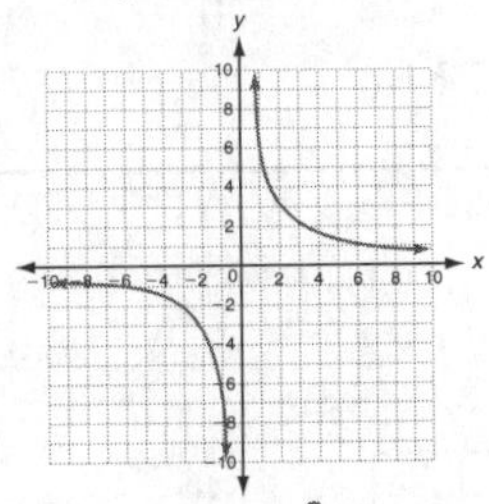

$y = \frac{6}{x}$

6. Write and graph the inverse variation in which $y = 1$ when $x = -3$.

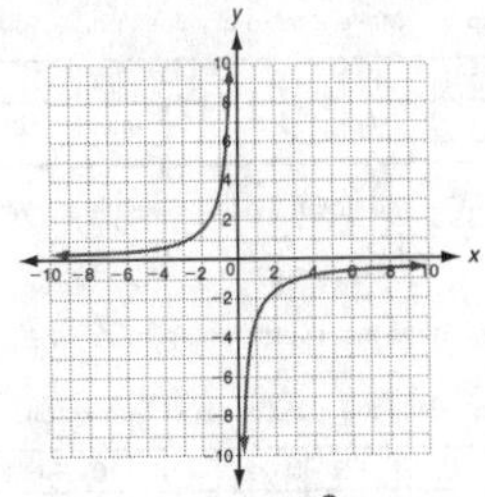

$y = \frac{-3}{x}$

7. Let $x_1 = 4$, $y_1 = 12$, and $x_2 = 3$. Let $y$ vary inversely as $x$. Find $y_2$. 16

8. Let $x_1 = 3$, $y_1 = 10$, and $y_2 = 15$. Let $y$ vary inversely as $x$. Find $x_2$. 2

9. While traveling in a car, the speed of the car is inversely proportional to the time it takes to travel a certain distance. At 25 mi/h, it takes 15 minutes to travel to work. How many minutes would it take traveling 30 mi/h? 12.5 minutes

10. The amount of pizza that Kirby can buy varies inversely as the price of the pizza increases. Kirby can afford to buy 3 pizzas that cost \$15.00 each. How many pizzas that cost \$9.00 each can Kirby buy? 5 pizzas




LESSON 12-1

## Practice C

### Inverse Variation

**Tell whether each relationship is an inverse relation. Explain.**

1.

| x | y |
|---|---|
| 0.25 | 3.5 |
| 0.5 | 1.4 |
| 0.75 | 0.875 |

no; $xy$ is not constant.

2.

| x | y |
|---|---|
| 1 | 3.5 |
| 2.5 | 1.4 |
| 4 | 0.875 |

yes; $xy$ is constant.

3. $2x + 4y = 6$

no; the equation cannot be written in the form $y = \frac{k}{x}$.

4. $3xy = 12$

yes; the equation can be written in the form $y = \frac{k}{x}$.

5. Write and graph the inverse variation in which $y = -4$ when $x = 2$.

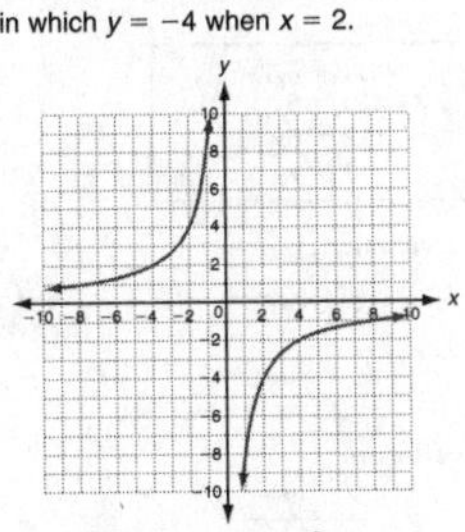

$y = \frac{-8}{x}$

6. Write and graph the inverse variation in which $y = \frac{1}{2}$ when $x = 36$.

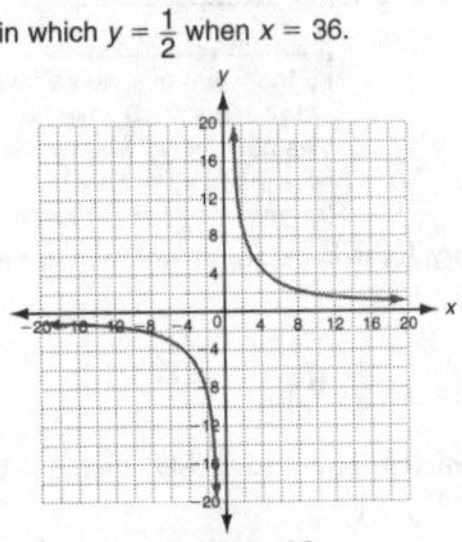

$y = \frac{18}{x}$

7. Let $x_1 = 8$, $y_1 = 2.5$, and $x_2 = 40$. Let $y$ vary inversely as $x$. Find $y_2$. 0.5

8. Let $x_1 = -1.5$, $y_1 = 4$, and $y_2 = 6$. Let $y$ vary inversely as $x$. Find $x_2$. −1

9. The number of hours $y$ it takes to paint a house varies inversely with the number of people painting $x$. It takes 2 people 20 hours and 30 minutes to paint a house. How many people would be needed if the house must be painted in 8 hours? 6 people

10. The volume of a gas $V$ varies inversely with the pressure $P$. If a gas with a volume of 30 L and a pressure of 2 atmospheres is compressed to a volume of 25 L, what will be the new pressure? 2.4 atmospheres




LESSON 12-1

## Reteach

### Inverse Variation

In an **inverse variation**, one quantity increases while the other quantity decreases. There are two ways to tell if a relationship is an inverse variation.

If the product of each ordered pair is constant, then the relationship is an inverse variation.

**Tell whether the relationship in the table is an inverse variation.**

Find $xy$ for each ordered pair.

| x | y | |
|---|---|---|
| 1 | 18 | *1 × 18 = 18* |
| 2 | 9 | *2 × 9 = 18* |
| 3 | 6 | *3 × 6 = 18* |

Because the product is a constant, the relationship is an inverse variation.

If you can write the function rule in $y = \frac{k}{x}$ form, the relationship is an inverse variation.

**Tell whether $xy = 4$ is an inverse variation.**

$xy = 4$

$\frac{xy}{x} = \frac{4}{x}$ *Divide both sides by x.*

$y = \frac{4}{x}$

This function is an inverse relation.

If $(x_1, y_1)$ and $(x_2, y_2)$ are solutions of an inverse variation, then $x_1y_1 = x_2y_2$.

**Let $x_1 = 4$, $y_1 = 3$, and $y_2 = 6$. Let $y$ vary inversely as $x$. Find $x_2$.**

$x_1y_1 = x_2y_2$ *Write the product rule for inverse variation.*

$(4)(3) = (x_2)(6)$ *Substitute $x_1 = 4$, $y_1 = 3$, and $y_2 = 6$.*

$12 = 6x_2$ *Multiply.*

$2 = x_2$ *Divide both sides by 6.*

**Tell whether each relationship is an inverse variation.**

1.

| x | y |
|---|---|
| 1 | 50 |
| 2 | 25 |
| 5 | 10 |

$xy =$ 50
$xy =$ 50
$xy =$ 50

yes

2.

| x | y |
|---|---|
| 2 | 12 |
| 3 | 8 |
| 4 | 6 |

$xy =$ 24
$xy =$ 24
$xy =$ 24

yes

3.

| x | y |
|---|---|
| 3 | 10 |
| 5 | 6 |
| 7 | 2 |

$xy =$ 30
$xy =$ 30
$xy =$ 14

no

4. $3x + y = 5$

no

5. $4xy = 8$

yes

6. $x = \frac{y}{10}$

no

7. Let $x_1 = 5$, $y_1 = 12$, and $y_2 = 6$. Let $y$ vary inversely as $x$. Find $x_2$. 10

8. Let $x_1 = 2$, $y_1 = 15$, and $x_2 = 10$. Let $y$ vary inversely as $x$. Find $y_2$. 3

LESSON 12-1
## Reteach
### Inverse Variation (continued)

You can also identify an inverse relation by its graph.

The graph of an inverse relation:
- has two parts that are not connected.
- does not contain (0, 0).

**Write and graph the inverse variation in which $y = 5$ when $x = 2$.**

**Step 1:** Find $k$.

$k = xy$

$= (2)(5)$

$= 10$

**Step 2:** Write the inverse variation equation.

$y = \frac{k}{x}$

$y = \frac{10}{x}$

**Step 3:** Make a table of values and graph.

| $x$ | −10 | −5 | −1 | 0 | 1 | 5 | 10 |
|---|---|---|---|---|---|---|---|
| $y$ | −1 | −2 | −10 | und. | 10 | 2 | 1 |

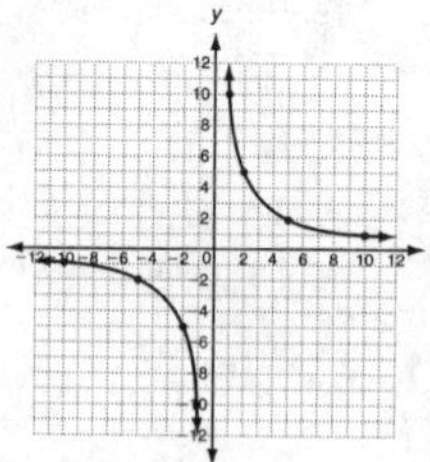

9. Write and graph the inverse variation when $y = 3$ and $x = 2$.

a. Find $k$. **6**

b. Write the inverse variation. $y = \frac{6}{x}$

c. Fill in the table of values and graph.

| $x$ | −3 | −2 | −1 | 0 | 1 | 2 | 3 |
|---|---|---|---|---|---|---|---|
| $y$ | −2 | −3 | −6 | und. | 6 | 3 | 2 |

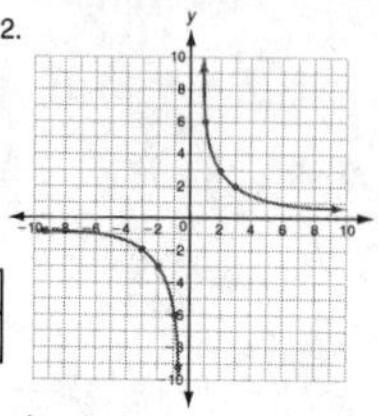

10. Write and graph the inverse variation when $y = 2$ and $x = -4$.

a. Find $k$. **−8**

b. Write the inverse variation. $y = \frac{-8}{x}$

c. Fill in the table of values and graph.

| $x$ | −4 | −2 | −1 | 0 | 1 | 2 | 4 |
|---|---|---|---|---|---|---|---|
| $y$ | 2 | 4 | 8 | und. | −8 | −4 | −2 |

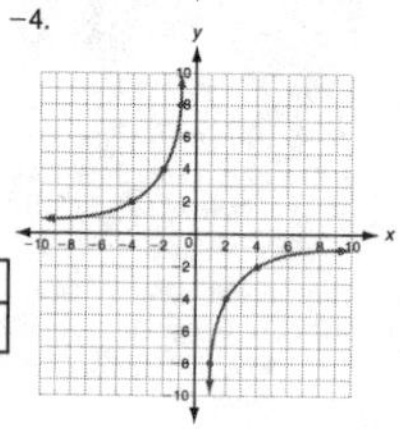




LESSON 12-1
## Challenge
### Seesaws

When two children of different weights sit on a seesaw that is joined to the pivot at its midpoint, one child sits higher than the pivot point, and the other child sits lower than it. They achieve balance and reverse who is higher and who is lower by exerting a downward force on the ground.

The diagram at right shows two different weights positioned at different distances from the pivot point of a seesaw. The equation indicates that weight and distance have an inverse variation. $w_1$ and $w_2$ represent weight, and $d_1$ and $d_2$ represent distance.

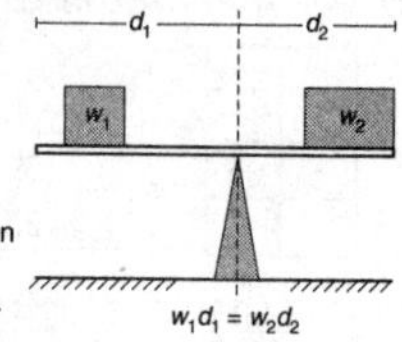


$w_1d_1 = w_2d_2$

**In Exercises 1–3, the total length of a seesaw is 12 feet, and boxes A and B are balanced.**

1. Boxes A and B weigh 60 pounds and 80 pounds, respectively. How far from the pivot point is box A if box B is 4 feet from the pivot point? **$5\frac{1}{3}$ feet**

2. Box A weighs 100 pounds and is 5 feet from the pivot point of the seesaw. How much does box B weigh if it is 6 feet from the pivot point? **$83\frac{1}{3}$ pounds**

3. Boxes A and B weigh 120 pounds and 150 pounds, respectively. How far is each box from the pivot point of the seesaw? (Hint: Let $x$ represent the distance between box A and the pivot point. Then write and solve an equation for $x$.) **Box A: $6\frac{2}{3}$ feet; Box B: $5\frac{1}{3}$ feet**

**In Exercises 4 and 5, solve each seesaw problem. The length of the board used to balance boxes A and B is 20 feet.**

4. Boxes A and B weigh 60 pounds and 80 pounds, respectively. John wants to put box A at the left end of the board and to position the pivot point 5 feet to the right of box A. How far from the right end of the board should he place box B? **$11\frac{1}{4}$ feet**

5. Boxes A and B weigh 60 pounds and 80 pounds, respectively. John wants to put box A at the left end of the board and to position the pivot point 15 feet to the right of box A. Is it possible to put box B on the seesaw so that the boxes balance? Explain your response.

**No; box B would have to be $11\frac{1}{4}$ feet from the pivot point. The board is not long enough.**

6. A board is $L$ feet long. Boxes A and B weigh 60 pounds and 80 pounds, respectively. Box A is placed $x$ feet from the pivot point.

a. Write and solve an equation for $L$ in terms of $x$. $L = \frac{7}{4}x$

b. Write and solve an equation for $x$ in terms of $L$. $x = \frac{4}{7}L$




LESSON 12-1
## Problem Solving
### Inverse Variation

**The number of gallons of gasoline that Isaac can afford $y$ varies inversely as the price per gallon $x$. Isaac can afford 12 gal when the price is \$2.50 per gallon.**

1. Write an inverse variation for this situation. **$xy = 30$ or $y = \frac{30}{x}$**

2. Determine a reasonable domain and range. **$x > 0$ and $y > 0$**

3. Complete this table of ordered pairs and graph the inverse variation.

| $x$ | 1.50 | 2.50 | 3.00 | 5.00 | 10.00 |
|---|---|---|---|---|---|
| $y$ | 20 | 12 | 10 | 6 | 3 |

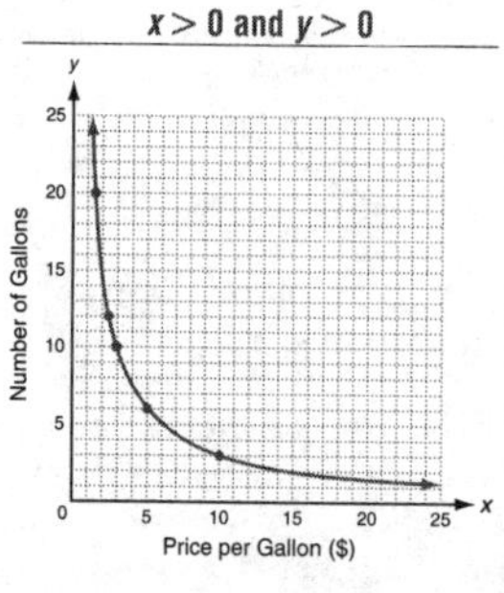


4. Use the graph to estimate the number of gallons Isaac can afford when the price is \$3.49 per gallon. **8.6 gal**

**Select the best answer.**

5. The inverse variation $xy = 9$ relates the current $x$ in amps to the resistance $y$ in ohms of a circuit attached to a 9-volt battery. What is the resistance of the circuit when the current is 1.8 amps?
   A 0.2 ohms C 10.8 ohms
   **(B) 5 ohms** D 16.2 ohms

6. Boyle's law states that the pressure of a quantity of gas $x$ varies inversely as the volume of the gas $y$. The volume of air inside a bicycle pump is 6.1 in$^3$, and the pressure is 18.3 psi. Which equation represents this situation?
   F $y = 3x$ H $y = 111.63x$
   G $y = \frac{3}{x}$ **(J) $y = \frac{111.63}{x}$**

7. The time it takes to drive between two cities varies inversely as speed. Dolores can drive between Houston, Texas, and San Antonio, Texas, in 3.5 h at 60 mi/h. How long will the trip take if she drives 70 mi/h? (Round your answer to the nearest tenth.)
   A 2.3 h **(C) 3 h**
   B 2.6 h D 4.1 h

8. On a balanced lever, weight varies inversely as the distance from the fulcrum. Angie weighs 120 lb. When she sits $2\frac{1}{2}$ ft from the fulcrum of a seesaw, she balances with her brother, who weighs 80 lb. How far is her brother from the fulcrum of the seesaw?
   F $\frac{4}{15}$ ft H $1\frac{2}{3}$ ft
   G $\frac{3}{5}$ ft **(J) $3\frac{3}{4}$ ft**




LESSON 12-1
## Reading Strategies
### Analyzing Relationships

The diagram below explains what inverse variation is, how to identify it, and how to answer questions involving it.

An *inverse variation* is a relationship that can be written using the formula $y = \frac{k}{x}$, where $x \neq 0$ and $k$ is the *constant of variation.*

$y = \frac{k}{x}$

The relationship shown in the table

| $x$ | −2 | 1 | 3 |
|---|---|---|---|
| $y$ | −3 | 6 | 2 |

is an inverse variation because $k$, or $x$ times $y$, is the same (6) for each ordered pair.

The equation $y = \frac{4}{x}$ is an inverse variation because it is in the form $y = \frac{k}{x}$, where $k = 4$.

If the value of $y$ varies inversely as $x$ and $y_1 = 3$ when $x_1 = 4$, we can find $y_2$ when $x_2 = 6$ by solving $x_1y_1 = x_2y_2$.

$x_1y_1 = x_2y_2$
$(4)(3) = 6y_2$
$12 = 6y_2$
$2 = y_2$

**Identify $k$, the constant of variation, for each inverse variation.**

1. $y = \frac{5}{x}$ **5**
2. $y = -\frac{1}{x}$ **−1**
3. $xy = -7$ **−7**

**Tell whether or not each relationship is an inverse variation.**

4. $y + 9 = x$ **no**

5. $y = \frac{3}{x}$ **yes**

6.

| $x$ | −4 | −2 | −1 |
|---|---|---|---|
| $y$ | −2 | −4 | −8 |

**yes**

7.

| $x$ | 1 | 2 | 3 |
|---|---|---|---|
| $y$ | 3 | 2 | 1 |

**no**

**Let $y$ vary inversely as $x$. Find each indicated value.**

8. $x_1 = 1$, $y_1 = -9$, $y_2 = 3$; find $x_2$ **−3**

9. $x_1 = 4$, $y_1 = 5$, $x_2 = -2$; find $y_2$ **−10**

10. $x_1 = 3$, $y_1 = 5$, $y_2 = 1$; find $x_2$ **15**

11. $x_1 = 2$, $y_1 = -4$, $x_2 = -4$; find $y^2$ **2**

LESSON 12-2

# Practice A

## Rational Functions

**Identify the excluded value for each rational function.**

1. $y = \frac{2}{x}$ **0**

2. $y = \frac{2}{x+1}$ **−1**

3. $y = \frac{5}{x-7}$ **7**

**Identify the asymptotes.**

4. $y = \frac{1}{x-6}$ **$x = 6$; $y = 0$**

5. $y = \frac{2}{x+4} - 5$ **$x = -4$; $y = -5$**

6. $y = \frac{3}{x-1} + 4$ **$x = 1$; $y = 4$**

**Graph each function.**

7. $y = \frac{8}{x+2}$

a. Vertical asymptote: **$x = -2$**

b. Horizontal asymptote: **$y = 0$**

c. Graph.

| x | y |
|---|---|
| −10 | −1 |
| −6 | −2 |
| −4 | −4 |
| 0 | 4 |
| 2 | 2 |
| 6 | 1 |

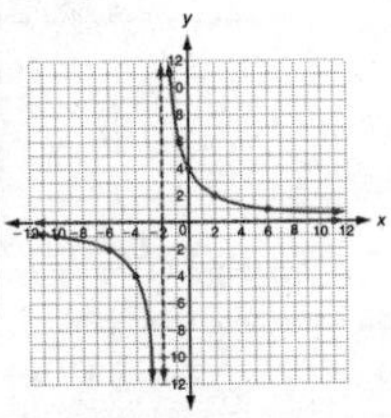

8. $y = \frac{20}{x+5} - 5$

a. Vertical asymptote: **$x = -5$**

b. Horizontal asymptote: **$y = -5$**

c. Graph.

| x | y |
|---|---|
| −25 | −6 |
| −15 | −7 |
| −10 | −9 |
| 0 | −1 |
| 5 | −3 |
| 15 | −4 |

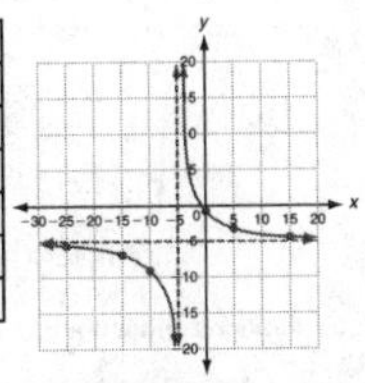

9. A caterer has \$50 in his budget for cheddar cheese. If $x$ represents the cost per pound, then $y = \frac{50}{x}$ represents the number of pounds $y$ he can buy.

a. Describe the reasonable domain and range values.

**D: nonnegative values**

**R: nonnegative values**

b. Graph the function.

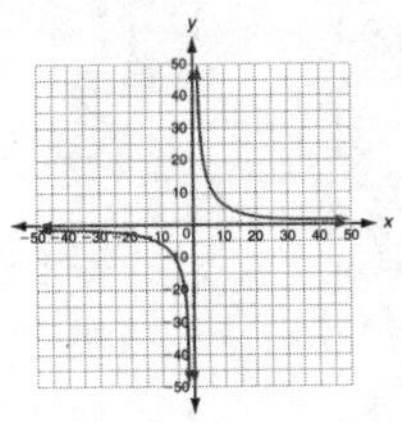





LESSON 12-2

# Practice B

## Rational Functions

**Identify the excluded value for each rational function.**

1. $y = \frac{-6}{x}$ **0**

2. $y = \frac{8}{x+3}$ **−3**

3. $y = \frac{5}{3x-6}$ **2**

**Identify the asymptotes.**

4. $y = \frac{5}{2x-5}$ **$x = \frac{5}{2}$; $y = 0$**

5. $y = \frac{2}{x+3} - 4$ **$x = -3$; $y = -4$**

6. $y = \frac{4}{x-6} + 3$ **$x = 6$; $y = 3$**

**Graph each function.**

7. $y = \frac{12}{x+3}$

a. Vertical asymptote: **$x = -3$**

b. Horizontal asymptote: **$y = 0$**

c. Graph.

| x | y |
|---|---|
| −9 | −2 |
| −7 | −3 |
| 0 | 4 |
| 1 | 3 |
| 3 | 2 |
| 9 | 1 |

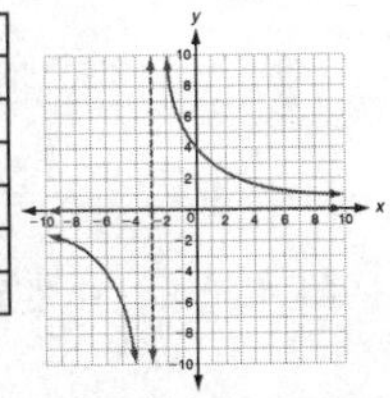

8. $y = \frac{6}{x-1} - 3$

a. Vertical asymptote: **$x = 1$**

b. Horizontal asymptote: **$y = -3$**

c. Graph.

| x | y |
|---|---|
| 7 | −2 |
| 4 | −1 |
| 0 | −9 |
| −1 | −6 |
| −2 | −5 |
| −5 | −4 |

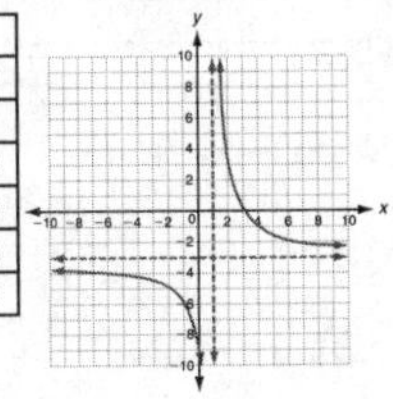

9. A music website is offering 5 free songs when you download any songs from their website. Catrina has \$25 to spend on songs. The number of songs $y$ that she can buy is given by $y = \frac{25}{x} + 5$, where $x$ is the price per song.

a. Describe the reasonable domain and range values.

**D: nonnegative values**

**R: whole numbers > 5**

b. Graph the function.

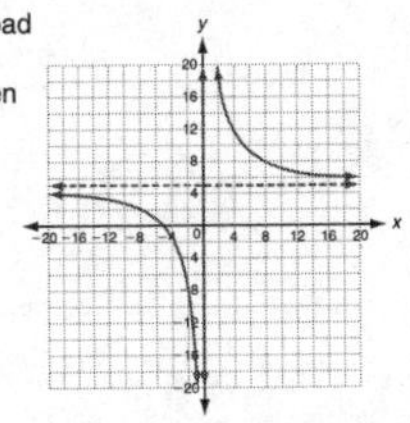





LESSON 12-2

# Practice C

## Rational Functions

**Identify the excluded value for each rational function.**

1. $y = \frac{4}{3x}$ **0**

2. $y = \frac{2}{4x+3}$ **$-\frac{3}{4}$**

3. $y = \frac{5}{6x-1}$ **$\frac{1}{6}$**

**Identify the asymptotes.**

4. $y = \frac{1}{x-0.8}$ **$x = 0.8$; $y = 0$**

5. $y = \frac{6}{5x-1} + 2$ **$x = \frac{1}{5}$; $y = 2$**

6. $y = \frac{2}{0.5x+3} - 7$ **$x = -6$; $y = -7$**

**Graph each function.**

7. $y = \frac{3}{x-4}$

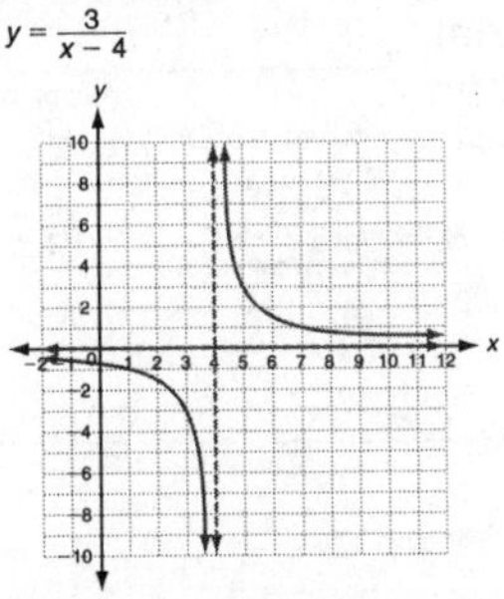

8. $y = \frac{10}{x+2} - 4$

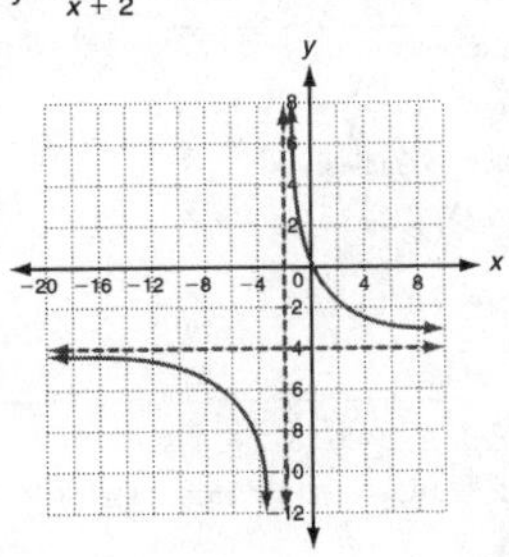

9. A furniture store is offering a payment plan which includes a finance charge of \$15 per month. The time in months $y$ it will take to pay of a bill of \$450, when $x$ dollars are paid each month, is $y = \frac{450}{x-15}$.

a. Describe the reasonable domain and range values.

**D: real numbers > 15**

**R: whole numbers > 0**

b. Graph the function.

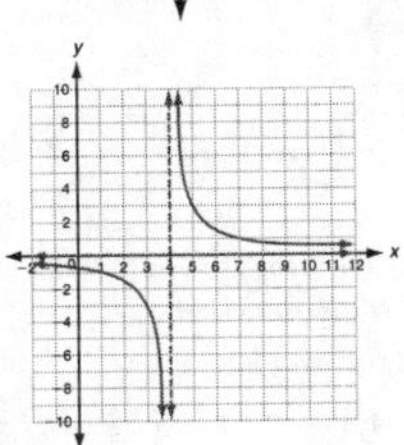

10. Write a rational function whose graph is the same shape as the graph of $f(x) = \frac{1}{x}$, but is translated 3 units right and 6 units up. **$y = \frac{1}{x-3} + 6$**





LESSON 12-2

# Reteach

## Rational Functions

Remember that division by zero is undefined. Because rational functions have $x$ in the denominator, we must exclude any values that make the denominator equal to zero.

**Identify the excluded value for $y = \frac{3}{x-2}$.**

The function will be undefined when $x - 2 = 0$.

$x - 2 = 0$

$\underline{\;+2 \quad +2\;}$

$x = 2$ The excluded value is 2.

For rational functions, a vertical asymptote will occur at excluded values. An asymptote is a line that a graph gets close to, but never touches. Most rational functions have a vertical and a horizontal asymptote.

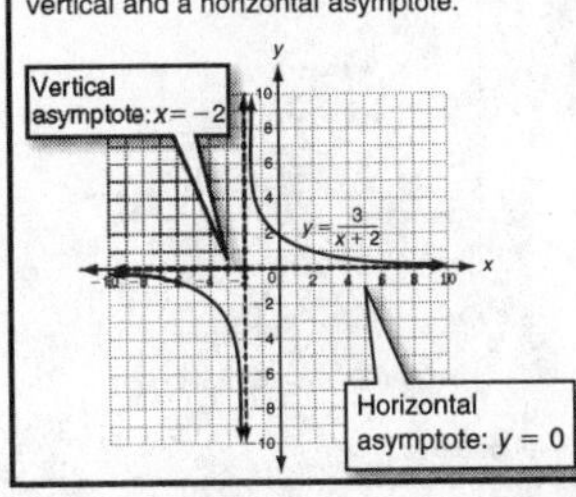



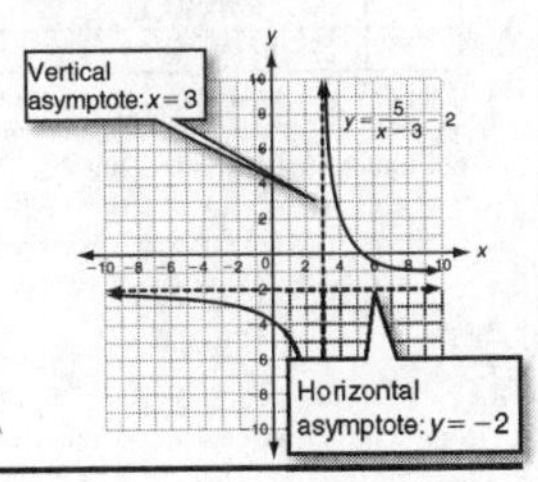



**Identify the excluded value for each rational function.**

1. $y = \frac{5}{x}$ **0**

2. $y = -\frac{6}{x-5}$ **5**

3. $y = \frac{4}{x+7}$ **−7**

**State the asymptotes for each graph below.**

4. 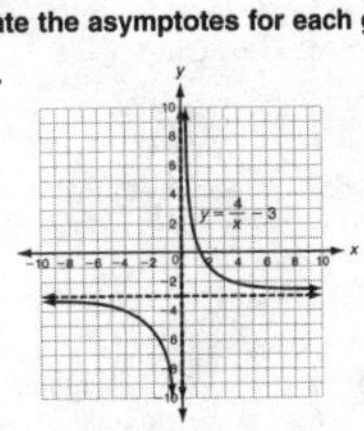

**$x = 0$, $y = -3$**

5. 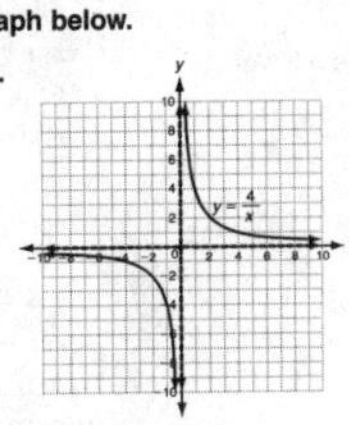

**$x = 0$, $y = 0$**

6. 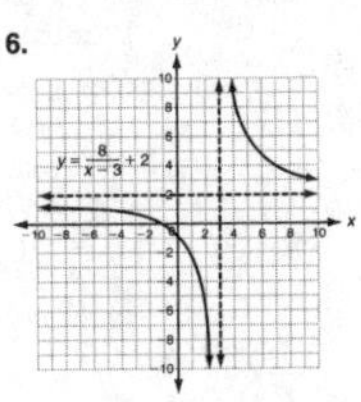

**$x = 3$, $y = 2$**

LESSON 12-2

## Reteach

### Rational Functions (continued)

You can also determine the asymptotes from the rational function itself.

$$y = \frac{a}{x+b} + c$$

- vertical asymptote at $x = -b$
- horizontal asymptote at $y = c$

You can use the asymptotes to help you graph.

**Graph** $y = \frac{12}{x+2}$.

**Step 1:** Identify the vertical and horizontal asymptotes.

$b = 2$, so the vertical asymptote is $x = -2$.

$c = 0$, so the horizontal asymptote is $y = 0$.

**Step 2:** Graph the asymptotes.

**Step 3:** Make a table of values.

Choose $x$ values on both sides of the vertical asymptote.

| $x$ | −8 | −6 | −4 | 1 | 2 | 4 |
|---|---|---|---|---|---|---|
| $y$ | −2 | −3 | −6 | 4 | 3 | 2 |

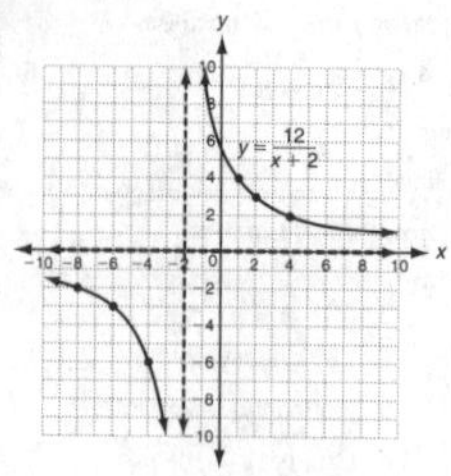

**Identify the asymptotes.**

7. $y = \frac{4}{x+5}$ $x = -5, y = 0$

8. $y = \frac{3}{x+2} - 6$ $x = -2, y = -6$

9. $y = \frac{1}{x-3} + 8$ $x = 3, y = 8$

**Graph each function.**

10. $y = \frac{4}{x-1}$

asymptotes: $x = 1, y = 0$

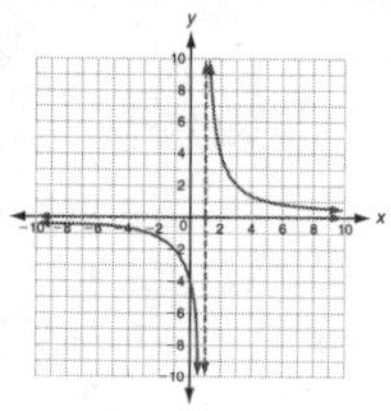

11. $y = \frac{8}{x+2} + 3$

asymptotes: $x = -2, y = 3$

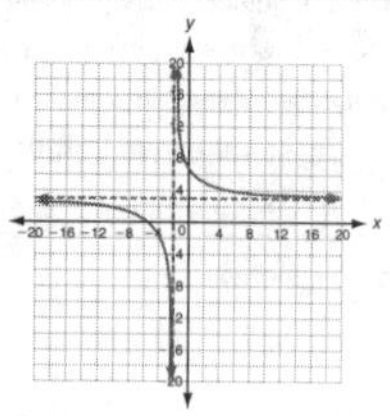





LESSON 12-2

## Challenge

### Exploring Functions Defined by Reciprocals

The function $f(x) = x$ is a polynomial function. The function $g(x) = \frac{1}{x}$ is a rational function. The diagram at right shows the graphs of $f(x) = x$ and $g(x) = \frac{1}{x}$.

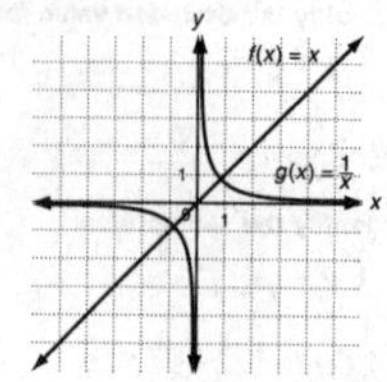

Notice that the graph of $f$ is a continuous and straight line. The graph of $g$ is neither continuous nor straight. Taking the reciprocal of $x$ has the effect of breaking a straight line into two curved branches, each extending indefinitely. Notice also that the break occurs at $x = 0$, the only real number that has no reciprocal.

**In Exercises 1 and 2, graph $f$ and $g$ on the same coordinate grid. For what value of $x$ does the break occur?**

1. $f(x) = x + 1$ and $g(x) = \frac{1}{x+1}$

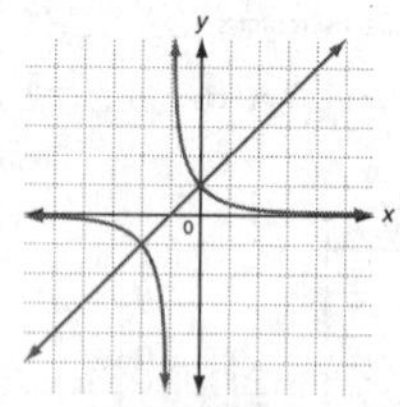

break at $x = -1$

2. $f(x) = x - 1$ and $g(x) = \frac{1}{x-1}$

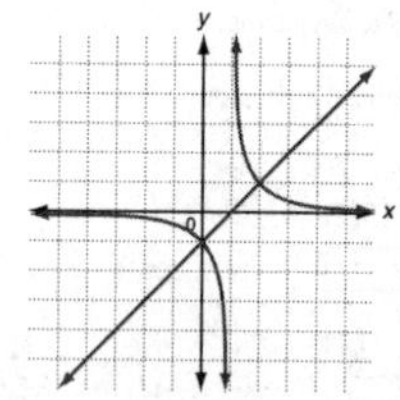

break at $x = 1$

**Graph each function and its reciprocal on the same coordinate grid.**

3. $y = x - 2.5$

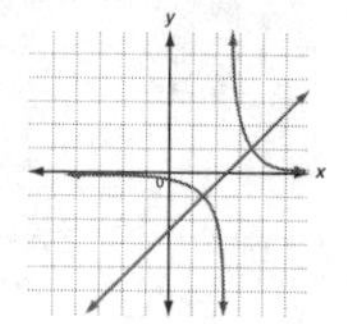

$y = x - 2.5$ and $y = \frac{1}{x - 2.5}$

4. $y = x - 3$

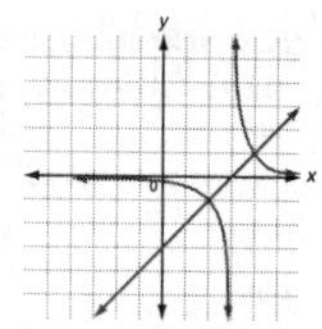

$y = x - 3$ and $y = \frac{1}{x-3}$

5. $y = x + 3$

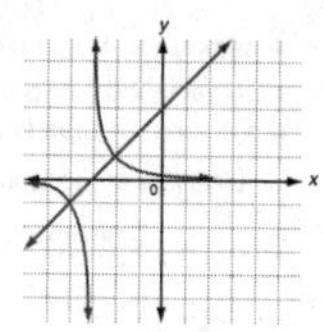

$y = x + 3$ and $y = \frac{1}{x+3}$





LESSON 12-2

## Problem Solving

### Rational Functions

**Gabrielle saves \$100 and decides to join a CD Club. As a member, she gets 12 CDs free and then must buy the rest at the regular price of $x$ dollars. The number of CDs $y$ she can buy is $y = \frac{100}{x} + 12$.**

1. Determine a reasonable domain and range.

D: $x > 0$; R: natural numbers greater than 12

2. Identify the vertical and horizontal asymptotes.

$x = 0$ and $y = 12$

3. Complete this table of ordered pairs and graph the rational function.

| $x$ | 5 | 10 | 20 | 25 |
|---|---|---|---|---|
| $y$ | 32 | 22 | 17 | 16 |

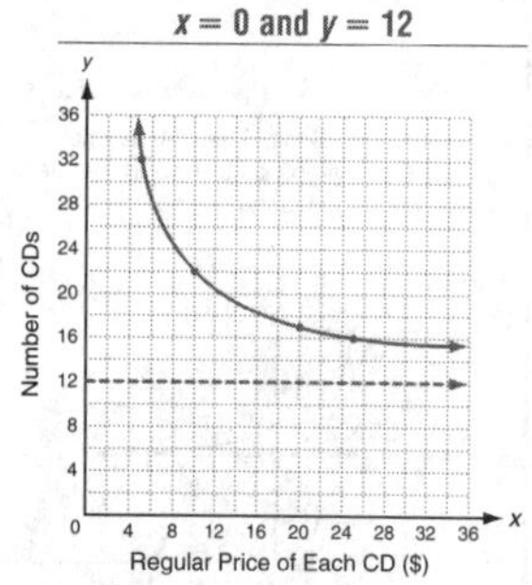

4. Gabrielle learns that the regular price of a CD is \$17.90. How many CDs can she buy through the Club with \$100?

17

**Select the best answer. For 5–6, refer to the situation above.**

5. Before joining the CD Club, Gabrielle learns about the Student Savings Card. If she buys a card for \$20, she gets \$5 off every CD she buys at the local record store. If each CD costs $x$ dollars at the local record store, then the number of CDs $y$ she can buy is $y = \frac{100-20}{x-5}$. The average price of each CD at the local record store is \$14.95. How many CDs can she buy with the Card and \$100?

A 4 
Ⓑ 8 
C 10 
D 12

6. Gabrielle realizes she must pay \$2.95 for shipping and handling on every CD from the club, including the 12 free CDs. That changes the function for the club to $y = \frac{100 - 2.95(12)}{x + 2.95} + 12$. If the regular price is still \$17.90, how many CDs can she buy through the Club?

F 3 
Ⓖ 15 
H 16 
J 18

7. At $x$ meters away from a very loud stereo speaker, the intensity of the sound $y$ in Watts per square meter is given by $y = \frac{0.0016}{x^2}$. What is the intensity when you stand 4 meters away?

Ⓐ 0.0001 W/m$^2$ 
B 0.0002 W/m$^2$ 
C 0.0004 W/m$^2$ 
D 0.0008 W/m$^2$





LESSON 12-2

## Reading Strategies

### Use a Concept Map

The concept map below will help you understand rational functions.

**Rational Function**

**Definitions**

A *rational function* is a function whose rule has a variable in the denominator.

An *excluded value* for a rational function is any value that makes the denominator equal to 0.

**Graph**

$y = \frac{1}{x-1} + 2$

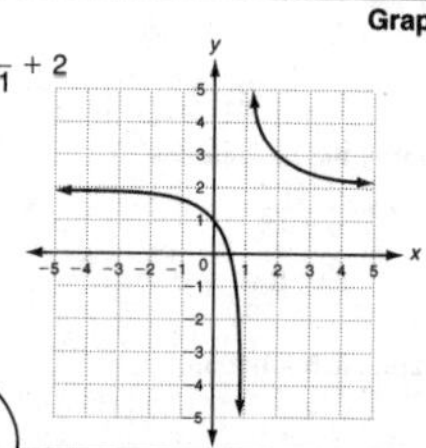

**Examples**

$y = \frac{4}{x}$; $y = \frac{1}{x^2}$; $y = \frac{2}{x-5} + 3$

**Non-Examples**

$y = \frac{x}{6}$; $y = x + \frac{1}{2}$; $y = 3x^2$

**Asymptotes**

A rational function in the form $y = \frac{a}{x+b} + c$ has a *vertical asymptote* at the excluded value $x = -b$, and a *horizontal asymptote* at $y = c$.

**Complete the following.**

1. Tell whether or not each of the following is a rational function.

a. $y = \frac{x}{3} + 2$ no

b. $y = \frac{3}{x^2} - 1$ yes

c. $y = \frac{1}{x+8}$ yes

2. Explain why $y = x + \frac{1}{2}$ is not a rational function.

It has no variable in the denominator.

3. Describe how to find the excluded value of the rational function $y = \frac{3}{x-5}$. Then find it.

Find what value makes $x - 5 = 0$; 5

4. Identify the asymptotes for the rational function $y = \frac{1}{x-1} + 2$ graphed above.

vertical: $x = 1$

horizontal: $y = 2$

LESSON 12-3

## Practice A
### Simplifying Rational Expressions

**Find any excluded values for each rational expression.**

1. $\frac{2}{x-5}$ 　 5

2. $\frac{4}{x^2+3x}$ 　 0; −3

3. $\frac{x+3}{x^2+6x+8}$ 　 −4, −2

**Simplify each rational expression, if possible.**

4. $\frac{x+4}{x^2-16}$ 　 $\frac{x+4}{(x+4)(x-4)}$ 　 $\frac{1}{x-4}$

5. $\frac{3x-18}{x^2-36}$ 　 $\frac{3(x-6)}{(x+6)(x-6)}$ 　 $\frac{3}{x+6}$

6. $\frac{x^2-1}{x^2+8x+7}$ 　 $\frac{(x+1)(x-1)}{(x+1)(x+7)}$ 　 $\frac{x-1}{x+7}$

7. $\frac{x+5}{x^2+7x+10}$ 　 $\frac{1}{x+2}$

8. $\frac{x^2+4x-5}{x^2-4x+3}$ 　 $\frac{x+5}{x-3}$

9. $\frac{x^2-2x}{x^2+2x-8}$ 　 $\frac{x}{x+4}$

10. $\frac{3-x}{2x-6}$ 　 $\frac{-1(x-3)}{2(x-3)}$ 　 $-\frac{1}{2}$

11. $\frac{x-4}{16-x^2}$ 　 $-\frac{1}{x+4}$

12. $\frac{8-x}{x^2-6x-16}$ 　 $-\frac{1}{x+2}$

13. An animal's surface-area-to-volume ratio indicates the amount of heat loss the animal may experience. Because a smaller animal has a higher ratio of surface-area-to-volume, the smaller animal has a higher rate of heat loss.

a. Find the surface-area-to-volume ratio of a cube. (*Hint*: For a cube, $S = 6s$ and $V = s^3$.) 　 $\frac{6}{s^2}$

b. Two animals are both close to a cube in shape. The height of the first animal is 4 cm. The height of the second animal is 6 cm. Which animal would experience the greater amount of heat loss? Explain.

ratio for first animal: $\frac{3}{8}$; ratio for second animal: $\frac{1}{6}$; the first animal would experience greater heat loss because it's surface-area-to-volume ratio is greater.




LESSON 12-3

## Practice B
### Simplifying Rational Expressions

**Find any excluded values for each rational expression.**

1. $\frac{6}{3+x}$ 　 −3

2. $\frac{5}{x^2-4x}$ 　 0; 4

3. $\frac{x+6}{x^2+3x-4}$ 　 −4, 1

**Simplify each rational expression, if possible. Identify any excluded values.**

4. $\frac{7}{x-3}$ 　 $\frac{7}{x-3}$; 3

5. $\frac{5x^2+10x}{5x}$ 　 $x+2$; 0

6. $\frac{2x}{4x^2+6x}$ 　 $\frac{1}{2x+3}$; 0, $-\frac{3}{2}$

**Simplify each rational expression, if possible.**

7. $\frac{x+3}{x^2-2x-15}$ 　 $\frac{1}{x-5}$

8. $\frac{3x+6}{x^2+3x+2}$ 　 $\frac{3}{x+1}$

9. $\frac{x-6}{x^2-7x+6}$ 　 $\frac{1}{x-1}$

10. $\frac{x^2-49}{x^2+8x+7}$ 　 $\frac{x-7}{x+1}$

11. $\frac{x^2+4x-5}{x^2-4x+3}$ 　 $\frac{x+5}{x-3}$

12. $\frac{x^2-2x}{x^2+2x-8}$ 　 $\frac{x}{x+4}$

13. $\frac{x^2-x-12}{4-x}$ 　 $-(x+3)$

14. $\frac{5-5x}{x^2-1}$ 　 $-\frac{5}{x+1}$

15. $\frac{3-x}{x^2-6x+9}$ 　 $-\frac{1}{x-3}$

16. When packaging food, a company wants a package that uses the least amount of material to hold the greatest volume of product. Some containers with mixed nuts are in the shape of a right circular cylinder.

a. Find the surface-area-to-volume ratio of a right circular cylinder. (Hint: For a right circular cylinder, $S = 2\pi rh + 2\pi r^2$ and $V = \pi r^2 h$.) 　 $\frac{2(r+h)}{rh}$

b. Container A has a radius of 2 in. and a height of 5 in. Container B has a radius of 4 in. and a height of 8 in. Which container should the company choose? Explain.

ratio for A: $\frac{7}{5}$; ratio for B: $\frac{3}{4}$; The company wants the smallest surface-area-to-volume ratio, so it should choose container B.




LESSON 12-3

## Practice C
### Simplifying Rational Expressions

**Find any excluded values for each rational expression.**

1. $\frac{5}{4x-7}$ 　 $\frac{7}{4}$

2. $\frac{4x}{x^2-4x}$ 　 0; 4

3. $\frac{x-5}{x^2-25}$ 　 −5, 5

**Simplify each rational expression, if possible. Identify any excluded values.**

4. $\frac{x-1}{x^2-4}$ 　 $\frac{x-1}{x^2-4}$; 2, −2

5. $\frac{4x^2-20x}{2x-10}$ 　 $2x$; 5

6. $\frac{x+1}{2x^3+5x^2+3x}$ 　 $\frac{1}{x(2x+3)}$; 0, $-\frac{3}{2}$, −1

**Simplify each rational expression, if possible.**

7. $\frac{x^2+4x-5}{x^2+8x+15}$ 　 $\frac{x-1}{x+3}$

8. $\frac{3x-12}{6x^2-18x-24}$ 　 $\frac{1}{2(x+1)}$

9. $\frac{2x^2-32}{x^2+8x+16}$ 　 $\frac{2(x-4)}{x+4}$

10. $\frac{10x^2-360}{5x^2+30x}$ 　 $\frac{2(x-6)}{x}$

11. $\frac{x^2+2x-8}{x^3-2x^2-24x}$ 　 $\frac{x-2}{x(x-6)}$

12. $\frac{x^2-1}{2x^2-3x-5}$ 　 $\frac{x-1}{2x-5}$

13. $\frac{5-x}{2x^3-16x^2+30x}$ 　 $-\frac{1}{2x(x-3)}$

14. $\frac{x^2-25}{10+3x-x^2}$ 　 $-\frac{x+5}{x+2}$

15. $\frac{27-3x^2}{6x^2-36x+54}$ 　 $-\frac{x+3}{2(x-3)}$

16. Flavored crushed ice is served in a paper container in the shape of a right circular cone.

a. Find the surface-area-to-volume ratio of a right circular cone. (*Hint*: For a right circular cone, $S = \pi r^2 + \pi rs$ (where $s = \sqrt{r^2+h^2}$) and $V = \frac{1}{3}\pi r^2 h$.) 　 $\frac{3(r+s)}{rh}$

b. Container A has a radius of 2 in. and a height of 3 in. Container B has a radius of 3 in. and a height of 4 in. Which container uses the least amount of paper to hold the greatest amount of crushed ice? Explain.

ratio for A: ≈2.8; ratio for B: 2; The container with the smallest ratio will hold the most ice using the least amount of paper and that is container B.




LESSON 12-3

## Reteach
### Simplifying Rational Expressions

A **rational expression** is an algebraic expression whose numerator and denominator are polynomials.

Exclude any values from a rational expression that make the denominator equal zero.

**Find any excluded value of $\frac{6x}{x^2-5x}$.**

$\frac{6x}{x^2-5x}$

$x^2 - 5x = 0$ 　 *Set denominator = 0.*

$x(x-5) = 0$ 　 *Factor.*

$x = 0$ or $x - 5 = 0$ 　 *Zero Product Property*

$+5 \quad +5$

$x = 5$

**The excluded values are 0 and 5.**

**Simplify $\frac{x+4}{2x^2+8x}$, if possible. Identify any excluded values.**

$\frac{x+4}{2x^2+8x}$

$\frac{x+4}{2x(x+4)}$ 　 Factor the denominator.

Find excluded values here.

$\frac{\cancel{x+4}^1}{2x\,(\cancel{x+4})^1}$ 　 Divide out common factors.

$\frac{1}{2x}$ 　 Simplify.

Remember to find the excluded value from the original equation (not the simplified one). **The excluded values are 0 and −4.**

**Identify any excluded values.**

1. $\frac{3}{4x}$ 　 0

2. $\frac{5x}{3x^2+15x}$ 　 0; −5

3. $\frac{6}{x^2-5x-14}$ 　 −2, 7

**Simplify each rational expression, if possible. Identify any excluded values.**

4. $\frac{4x}{20x^3}$ 　 $\frac{1}{5x^2}$; $x \neq 0$

5. $\frac{x+3}{x-4}$ 　 $\frac{x+3}{x-4}$; $x \neq 4$

6. $\frac{27x}{3x^2}$ 　 $\frac{9}{x}$; $x \neq 0$

7. $\frac{6}{3x+9}$ 　 $\frac{2}{x+3}$; $x \neq -3$

8. $\frac{x+4}{2x^2+8x}$ 　 $\frac{1}{2x}$; $x \neq 0$, $x \neq -4$

9. $\frac{5x^2-20x}{x-4}$ 　 $5x$; $x \neq 4$

LESSON 12-3 **Reteach**

## Simplifying Rational Expressions (continued)

Rational expressions with binomials, trinomials and opposite binomials can also be simplified.

**Simplify $\frac{x^2-4}{x^2+9x+14}$, if possible.**

$\frac{x^2-4}{x^2+9x+14}$

$\frac{(x+2)(x-2)}{(x+2)(x+7)}$ *Factor.*

$\frac{\cancel{(x+2)}^1(x-2)}{\cancel{(x+2)}^1(x+7)}$ *Divide out the common factors.*

$\frac{x-2}{x+7}$ *Simplify.*

**Simplify $\frac{3-x}{x^2-7x+12}$, if possible.**

$\frac{3-x}{x^2-7x+12}$

$\frac{3-x}{(x-3)(x-4)}$ *Factor.*

$\frac{-1(x-3)}{(x-3)(x-4)}$ *Factor −1 out of the numerator.*

$\frac{-1\cancel{(x-3)}^1}{\cancel{(x-3)}^1(x-4)}$ *Divide out the common factors.*

$-\frac{1}{x-4}$ *Simplify.*

**Simplify each rational expression, if possible.**

10. $\frac{x+3}{x^2-9}$ — $\frac{(x+3)}{(x-3)(x+3)}$ — $\frac{1}{x-3}$

11. $\frac{x^2-4x}{x^2-x-12}$ — $\frac{x(x-4)}{(x-4)(x+3)}$ — $\frac{x}{x+3}$

12. $\frac{x^2-2x-15}{x^2+5x+6}$ — $\frac{(x+3)(x-5)}{(x+3)(x+2)}$ — $\frac{x-5}{x+2}$

13. $\frac{x-9}{x^2-81}$ — $\frac{1}{x+9}$

14. $\frac{x^2+7x+6}{x+1}$ — $x+6$

15. $\frac{x+3}{x^2+7x+12}$ — $\frac{1}{x+4}$

16. $\frac{3x+15}{x^2-25}$ — $\frac{3}{x-5}$

17. $\frac{x-5}{25-x^2}$ — $-\frac{1}{x+5}$

18. $\frac{4-x}{x^2-2x-8}$ — $-\frac{1}{x+2}$





LESSON 12-3 **Challenge**

## More About Graphing Rational Functions

A rational function in the form $y=\frac{P(x)}{Q(x)}$, where $P(x)$ and $Q(x)$ are polynomials, may have vertical asymptotes, holes, or horizontal asymptotes according to the following rules.

**Example:** $y=\frac{x-1}{(x-1)(x-2)}$

**Vertical Asymptotes**

If $(x-r)$ is a factor of $Q(x)$ a greater numberof times than it is a factor of $P(x)$, then a vertical asymptote occurs at $x=r$.

**Holes**

If $(x-r)$ is a factor of $Q(x)$ a fewer or equal number of times than it is a factor of $P(x)$, then a hole occurs in the graph at $x=r$.

**Horizontal Asymptotes**

If the degree of $P(x)$ is ...

- greater than the degree of $Q(x)$, then there is no horizontal asymptote.
- less than the degree of $Q(x)$, then a horizontal asymptote occurs at $y=0$.
- equal to the degree of $Q(x)$, then a horizontal asymptote occurs at $y=\frac{\text{leading coefficient of } P(x)}{\text{leading coefficient of } Q(x)}$.

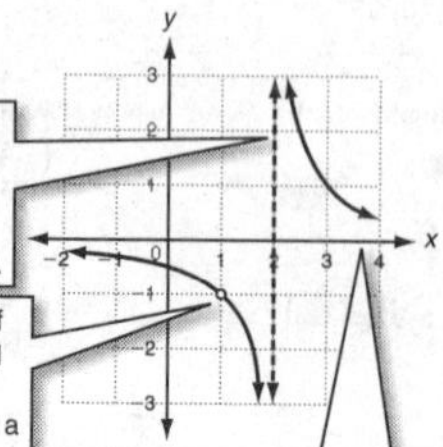

$(x-2)$ is a factor of the denominator once and is not a factor of the numerator. So a vertical asymptote occurs at $x=2$.

$(x-1)$ is a factor of the denominator and numerator an equal number of times. So a hole occurs at $x=1$.

The numerator has degree 1 and the denominator has degree 2. So a horizontal asymptote occurs at $y=0$.

**Graph these rational functions. First factor the numerator and denominator. Then use the rules above to identify asymptotes and holes. Finally, use a table of values to fill in-between.**

1. $y=\frac{3x^2-7x-6}{x^2-2x-3}=\frac{(3x+2)(x-3)}{(x+1)(x-3)}$

asymptotes: $x=-1$ and $y=3$

holes: at $x=3$

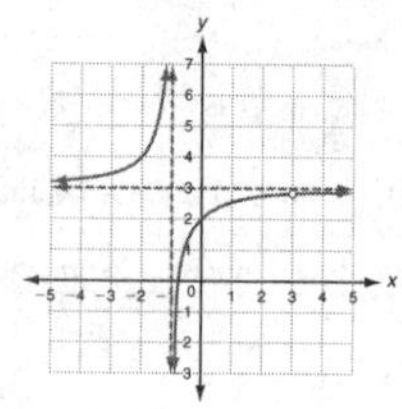

2. $y=\frac{x^2-3x-4}{-x^2+4}=\frac{(x+1)(x-4)}{-1(x+2)(x-2)}$

asymptotes: $x=2$, $x=-2$ and $y=-1$

holes: none

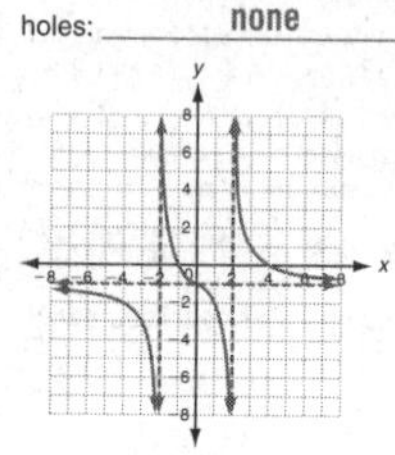





LESSON 12-3 **Problem Solving**

## Simplifying Rational Expressions

**A food company is considering two types of cylindrical cans for a new trail mix. The company wants a can that uses the least material for the greatest volume. For a cylinder, $S=2\pi r^2+2\pi rh$ and $V=\pi r^2h$.**

1. What is the surface-area-to-volume ratio for any cylinder?

$\frac{2(r+h)}{rh}$

2. Can A has radius 2 in. and height 3 in. What is the surface-area-to-volume ratio for Can A?

$\frac{10}{6}=1.\overline{6}$

3. Can B has radius 1.5 in. and height 4 in. What is the surface-area-to-volume ratio for Can B?

$\frac{11}{6}=1.8\overline{3}$

4. Which can should the company choose? Explain.

Can A; the ratio is lesser, Can A uses less material for each unit of volume.

**Select the best answer.**

5. For any circle with radius $r$, $P=2\pi r$ and $A=\pi r^2$. Which expression represents the perimeter-to-area ratio?

A $2r$

(B) $\frac{2}{r}$

C $\frac{r}{2}$

D $\frac{1}{2r}$

6. Tyrone takes an 8-by-8 square piece of cardboard and cuts $x$-by-$x$ squares out of each corner. (All units are inches.) He then folds up the sides to form a box with no top. Which expression represents the surface-area-to-volume ratio? (*Hint:* For the surface area, find the area of the entire piece of cardboard and subtract the squares that were removed.)

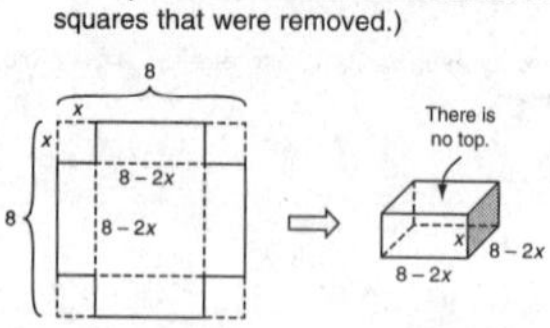


F $\frac{1}{x}$

G $\frac{1}{2x(1-16x)}$

H $\frac{8}{x(4-x)}$

(J) $\frac{4+x}{x(4-x)}$

7. In 2000, there are $x$ mountain lions (predators) and $x$ deer (prey) in a wildlife preserve. In 2005, the populations level off at $3x$ mountain lions and $4x$ deer, for a predator-to-prey ratio of $\frac{3}{4}$. If the wildlife preserve adds 12 more deer in 2007, how many mountain lions should also be added to keep the predator-to-prey ratio equivalent to $\frac{3}{4}$?

(A) 9

B 12

C 16

D 20





LESSON 12-3 **Reading Strategies**

## Use a Table

The table below shows examples of the many ways that rational expressions can be simplified.

| Use Properties of Exponents | Factor out Common Monomials | Factor Trinomials & Special Products | Use Opposite Binomials |
|---|---|---|---|
| $\frac{4m^5}{24m^2}$ | $\frac{2b^2-18b}{b-9}$ | $\frac{x^2-9}{x^2+2x-15}$ | $\frac{6t-t^2}{t^2-4t-12}$ |
| $\frac{m^3}{6}$ | $\frac{2b(b-9)}{b-9}$ | $\frac{(x+3)(x-3)}{(x+5)(x-3)}$ | $\frac{t(6-t)}{(t+2)(t-6)}$ |
| | $2b$ | $\frac{x+3}{x+5}$ | $\frac{t(-1)(t-6)}{(t+2)(t-6)}$ |
| | | | $-\frac{t}{(t+2)}$ |

**Use the table to answer the following.**

1. Look at the first column. Explain how $\frac{4m^5}{24m^2}$ was simplified to $\frac{m^3}{6}$.

$\frac{4}{24}$ simplified to $\frac{1}{6}$ and the exponents were subtracted, making $\frac{m^5}{m^2}=m^3$.

2. Look at the second column. What expression was divided out of the numerator and denominator? $b-9$

3. Look at the third column. What type of special product was in the numerator?

difference of squares

4. Look at the fourth column. What happened in the numerator?

$(6-t)$ became $(-1)(t-6)$

**Simplify each rational expression.**

5. $\frac{3t^4}{9t^6}$ — $\frac{1}{3t^2}$

6. $\frac{n^4-3n^3}{2n^2-6n}$ — $\frac{n^2}{2}$

7. $\frac{4c+4}{2c^2-5c-7}$ — $\frac{4}{2c-7}$

8. $\frac{18-2x}{x^2-81}$ — $-\frac{2}{x+9}$

LESSON 12-4
## Practice A
### Multiplying and Dividing Rational Expressions

**Multiply. Simplify your answer.**

1. $\frac{3x^2}{2y^4} \cdot \frac{4xy^4}{x^6}$ **$\frac{6}{x^3}$**

2. $\frac{2q-8}{6q^2} \cdot \frac{q^5}{pq-4p}$ **$\frac{q^3}{3p}$**

3. $\frac{ab}{5ab-5b} \cdot (a^2+3a-4)$ **$\frac{a^2+4a}{5}$**

4. $\frac{2}{4m^2+8m} \cdot \frac{m^2+8m+12}{6+m}$ **$\frac{1}{2m}$**

**Divide by multiplying the reciprocal. Simplify your answer.**

5. $\frac{c^4d}{9c} \div \frac{2cd^8}{3c^4d^2}$ **$\frac{c^6}{6d^5}$**

6. $\frac{k^2-6k+9}{k+1} \div \frac{8k^2-24k}{2k^2}$ **$\frac{k^2-3k}{4k+4}$**

7. For a game at school, Alma is asked to pull the names of two students out of a bag without looking, and without replacing the first name. There are 3 more boys than girls in the class.

   a. Let $x$ represent the number of girls in the class. Write an expression to represent the number of boys. **$x+3$**

   b. Write a simplified expression to represent the total number of students (girls + boys) in the class. **$2x+3$**

   c. Write and simplify an expression that represents the probability of Alma picking a girl, then a boy. *Hint*: Use $P(A \text{ and } B) = P(A) \cdot P(B \text{ after } A)$. **$\frac{x(x+3)}{(2x+3)(2x+2)}$**

   d. Substitute 10 for $x$ in your expression from part c to find the probability that Alma picks a girl then a boy when there are 10 girls in the class. Round to the nearest whole percent. **$\approx 26\%$**




LESSON 12-4
## Practice B
### Multiplying and Dividing Rational Expressions

**Multiply. Simplify your answer.**

1. $\frac{8a^2b^5}{a^3} \cdot \frac{3a^2}{4b^9}$ **$\frac{6a}{b^4}$**

2. $\frac{4x+8}{3} \cdot \frac{6x}{x+2}$ **$8x$**

3. $\frac{7}{2t-6} \cdot (t^2+t-12)$ **$\frac{7t+28}{2}$**

4. $\frac{3x^2+xy^3}{y^3} \cdot \frac{2xy+8y}{4x+x^2}$ **$\frac{6x+2y^3}{y^2}$**

**Divide. Simplify your answer.**

5. $\frac{5j^2k^2}{3jk^5} \div \frac{10j^2k}{9j^3}$ **$\frac{3j^2}{2k^4}$**

6. $\frac{3c^2+24c}{c^2-2c+1} \div \frac{c^2+9c+8}{9c-9}$ **$\frac{27c}{c^2-1}$**

7. Ramon is playing a game in which he must pull two blocks out of a bag containing red and yellow blocks. He cannot look, and he cannot replace the block. The bag has 4 more red blocks than yellow blocks.

   a. Write and simplify an expression that represents Ramon's probability of picking a red block, then a yellow block. **$\frac{x(x+4)}{2(x+2)(2x+3)}$**

   b. What is the probability that Ramon pulls a red block then a yellow block if there are 6 yellow blocks in the bag before his first pick? **25%**

   c. What is the probability that Ramon pulls two yellow blocks if there are 6 yellow blocks in the bag before his first pick? **12.5%**




LESSON 12-4
## Practice C
### Multiplying and Dividing Rational Expressions

**Multiply. Simplify your answer.**

1. $\frac{9rt^5}{5r^4} \cdot \frac{10r^2}{27t}$ **$\frac{2t^4}{3r}$**

2. $\frac{6a+12}{3a+a^2} \cdot \frac{a+3}{4a^2+8a}$ **$\frac{3}{2a^2}$**

3. $\frac{2m}{5m+20} \cdot (m^2-16)$ **$\frac{2m^2-8m}{5}$**

4. $\frac{p+3}{p^2-25} \cdot \frac{3p-15}{p^2+p-6}$ **$\frac{3}{p^2+3p-10}$**

**Divide. Simplify your answer.**

5. $\frac{x^4y}{3z^5} \div \frac{x^2z^3}{9y^2z}$ **$\frac{3x^2y^3}{z^7}$**

6. $\frac{8n^2-8}{10n+10n^2} \div (2n^3+6n^2-8n)$ **$\frac{2}{5n^3+20n^2}$**

7. At the town fair, Jillian is in charge of pulling two winning tickets out of a box containing tickets for a prize drawing. There are 12 more tickets from female entrants than from male entrants.

   a. If the same person cannot win twice, write and simplify an expression that represents the probability of Jillian picking a female winner, then a male winner. **$\frac{x(x+12)}{2(x+6)(2x+11)}$**

   b. If the same person can win twice, write and simplify an expression that represents the probability of Jillian picking a female winner, then a female winner. **$\frac{(x+12)^2}{4(x+6)^2}$**

   c. What is the probability that Jillian pulls two female winning tickets if the same person can win twice, and if there are 25 tickets from female entrants before her first pick? **$\approx 43\%$**




LESSON 12-4
## Reteach
### Multiplying and Dividing Rational Expressions

If $a$, $b$, $c$, and $d$ are nonzero polynomials, then $\frac{a}{b} \cdot \frac{c}{d} = \frac{ac}{bd}$.

You can make any expression rational by writing it with a denominator of 1.

**Multiply $\frac{x+1}{7} \cdot \frac{5}{6x+6}$. Simplify your answer.**

$\frac{x+1}{7} \cdot \frac{5}{6x+6}$

$\frac{5(x+1)}{7(6x+6)}$ *Multiply the numerators and the denominators.*

$\frac{5(x+1)}{7 \cdot 6(x+1)}$ *Factor.*

$\frac{5(\cancel{x+1})^1}{7 \cdot 6(\cancel{x+1})^1}$ *Simplify.*

$\frac{5}{42}$

**Multiply $(3x+12) \cdot \frac{2}{x^2-x-20}$. Simplify your answer.**

$(3x+12) \cdot \frac{2}{x^2-x-20}$

$\frac{3x+12}{1} \cdot \frac{2}{x^2-x-20}$ *Write as a rational expression.*

$\frac{3(x+4)}{1} \cdot \frac{2}{(x+4)(x-5)}$ *Factor.*

$\frac{3(\cancel{x+4})^1}{1} \cdot \frac{2}{(\cancel{x+4})^1(x-5)}$ *Simplify.*

$\frac{6}{x-5}$

**Multiply. Simplify your answer.**

1. $\frac{x+3}{5x} \cdot \frac{2}{4x+12}$

   $\frac{\boxed{x+3}}{\boxed{5x}} \cdot \frac{\boxed{2}}{\boxed{4}\boxed{x+3}}$

   **$\frac{1}{10x}$**

2. $\frac{3x}{x+3} \cdot \frac{x^2+5x+6}{x}$ **$3x+6$**

3. $\frac{4x-12}{6x} \cdot \frac{x+3}{x^2-9}$ **$\frac{2}{3x}$**

4. $(4x+24)\frac{5}{x^2-36}$ **$\frac{20}{x-6}$**

5. $(x+7)\frac{3x}{x^2+13x+42}$ **$\frac{3x}{x+6}$**

6. $(x^2-16)\frac{6}{x^2-x-12}$ **$\frac{6x+24}{x+3}$**

LESSON 12-4 **Reteach**

## Multiplying and Dividing Rational Expressions (continued)

If $a$, $b$, $c$, and $d$ are nonzero polynomials, then $\frac{a}{b} \div \frac{c}{d} = \frac{a}{b} \cdot \frac{d}{c} = \frac{ad}{bc}$.

**Divide $\frac{3}{x+2} \div \frac{5}{x+2}$. Simplify your answer.**

$\frac{3}{x+2} \div \frac{5}{x+2}$

$\frac{3}{x+2} \cdot \frac{x+2}{5}$ *Write as multiplication by the reciprocal.*

$\frac{3}{\cancel{x+2}^1} \cdot \frac{\cancel{x+2}^1}{5}$ *Divide out common factors.*

$\frac{3}{5}$ *Simplify.*

**Divide $\frac{2x-4}{x^2} \div (x-2)$. Simplify your answer.**

$\frac{2x-4}{x^2} \div (x-2)$

$\frac{2x-4}{x^2} \div \frac{x-2}{1}$ *Write as a rational expression.*

$\frac{2x-4}{x^2} \cdot \frac{1}{x-2}$ *Write reciprocal.*

$\frac{2(x-2)}{x^2} \cdot \frac{1}{x-2}$ *Factor.*

$\frac{2\cancel{(x-2)}^1}{x^2} \cdot \frac{1}{\cancel{x-2}^1}$ *Simplify.*

$\frac{2}{x^2}$

**Divide. Simplify your answer.**

7. $\frac{4x}{x^2} \div \frac{x^2}{7}$

$\frac{4x}{x^2} \cdot \frac{7}{x^2}$

$\frac{28}{x^3}$

8. $\frac{8x^2y^3}{2xy} \div \frac{xy}{3}$

$12y$

9. $\frac{x^2}{x-3} \div \frac{x+5}{x^2+2x-15}$

$x^2$

10. $\frac{x+4}{x-5} \div (x^2 - x - 20)$

$\frac{1}{x^2 - 10x + 25}$

11. $\frac{x^2-4x-5}{8} \div (x^2 + 2x + 1)$

$\frac{x-5}{8x+8}$

12. $\frac{x+10}{x^2} \div (x^2 - 100)$

$\frac{1}{x^3 - 10x^2}$





LESSON 12-4 **Challenge**

## Multiplication and Division Puzzles (Revisited)

**For 1 and 2, draw a path from Start to Finish, multiplying by each new expression along the way, to find a product equal to the goal. Use only vertical or horizontal moves, and the path cannot cross itself.**

1. Goal: Product of 1

| | | |
|---|---|---|
| $\frac{x^2-x-6}{x+1}$ Start | $\frac{x^2+2x}{x-3}$ | $\frac{x^2-4}{x-3}$ |
| $\frac{x}{x+2}$ | $\frac{x^2+2x+1}{x^2}$ | $\frac{x+3}{x-1}$ |
| $\frac{x^2-9}{x^2-4}$ | $\frac{x}{x^2+3x+2}$ | $\frac{1}{x+2}$ Finish |

2. Goal: Product of $3(x-4)$

| | | |
|---|---|---|
| $\frac{18x^2-9x}{3x+2}$ | $\frac{3x^2+20x+12}{x^3-4x^2}$ | $\frac{27x^2}{6x}$ |
| $\frac{3x^2-10x-8}{6x^2+33x-18}$ Start | $\frac{6x^2+24x}{18x-9}$ | $\frac{6x-3}{6x+4}$ Finish |
| $\frac{2x^2-11x-6}{3x+2}$ | $\frac{x^2-8x+16}{4x^2-4x+1}$ | $\frac{6x^2+24}{3x-2}$ |

**For 3 and 4, draw a path from Start to Finish, dividing by each new expression along the way, to find a quotient equal to the goal. Use only vertical or horizontal moves, and the path cannot cross itself.**

3. Goal: Quotient of $\frac{15}{x}$

| | | |
|---|---|---|
| $\frac{x^2}{2y}$ | $\frac{1}{15y}$ | $6y^2$ Finish |
| $\frac{y}{10}$ | $\frac{x}{5y}$ | $\frac{5}{6}$ |
| $\frac{2}{3x}$ Start | $\frac{xy}{3}$ | $\frac{2}{x^3y}$ |

4. Goal: Quotient of 1

| | | |
|---|---|---|
| $\frac{x-3}{3x^2-10x+3}$ | $\frac{12x^2+11x-5}{x-3}$ Start | $\frac{x^2+7x-30}{3x^2+5x-2}$ |
| $\frac{x+2}{4x^2+45x+50}$ | $\frac{3x^2+5x-2}{x^2+7x-30}$ | $\frac{4x^2-7x-15}{x^2+12x+20}$ |
| $\frac{3x^2+29x-10}{x+10}$ | $\frac{4x^2+45x+50}{x+2}$ Finish | $\frac{x-3}{12x^2+11x-5}$ |





LESSON 12-4 **Problem Solving**

## Multiplying and Dividing Rational Expressions

**A bag contains 5 more red marbles than blue marbles. Janis is going to pick two marbles out of the bag without looking, and without replacing the first marble. Let $x$ represent the number of blue marbles.**

1. Write and simplify an expression that represents the probability that Janis will pick two blue marbles.

$\frac{x(x-1)}{(2x+5)(2x+4)}$

2. Is the probability that Janis picks a red marble and then a blue marble the same as the probability that she picks a blue and then a red? Explain.

Yes; $P$(red then blue) $= \frac{x+5}{2x+5} \cdot \frac{x}{2x+4}$ and $P$(blue then red) $= \frac{x}{2x+5} \cdot \frac{x+5}{2x+4}$; both $= \frac{x(x+5)}{(2x+5)(2x+4)}$.

3. Write and simplify an expression that represents the probability that Janis will pick two red marbles. Then find the probability of picking two red marbles if the bag contains 20 blue marbles.

$\frac{(x+5)(x+4)}{(2x+5)(2x+4)}$; $\frac{10}{33} \approx 0.3$

**A square dart board has three regions, A, B, and C, as shown. Units are in inches. Select the best answer.**

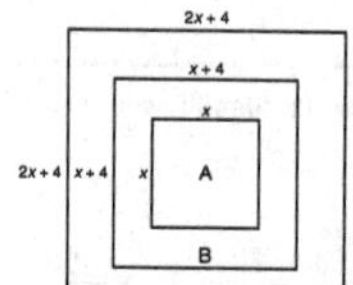


4. Which expression represents the probability of throwing two darts that land in region A and then region B?

A $\frac{2x^2}{x+2}$

(B) $\frac{x^2}{2(x+2)^3}$

C $\frac{(x+4)^2}{16(x+2)^4}$

D $\frac{x^2(x+4)^2}{4(x+2)^2}$

5. Divide two rational expressions to find how many times greater the probability of a dart landing in region C is than the probability of a dart landing in region A.

(F) $\frac{3x+8}{x}$

G $\frac{x}{2(x+1)}$

H $\frac{4(x+2)^2}{x^2}$

J $\frac{x(3x+8)}{8(x+2)}$

6. Which expression represents the probability of throwing two darts that both land in region B?

(A) $\frac{4}{(x+2)^2}$

B $\frac{x^2}{(x+4)^2}$

C $\frac{(x+4)^4}{16(x+2)^4}$

D $\frac{64(x+2)^2}{x^2(3x+8)^2}$

7. If $x = 2$ in., what is the probability of throwing two darts that both land in region A?

F $\frac{1}{4}$

G $\frac{1}{16}$

H $\frac{1}{32}$

(J) $\frac{1}{256}$





LESSON 12-4 **Reading Strategies**

## Connecting Concepts

The rules for multiplying and dividing rational expressions are the same as the rules for multiplying and dividing fractions. Look at the connections below.

FRACTIONS | RATIONAL EXPRESSIONS

Rule 1: $\frac{a}{b} \cdot \frac{c}{d} = \frac{ac}{bd}$

$\frac{8}{9} \cdot \frac{3}{5} = \frac{8 \cdot 3}{9 \cdot 5} = \ldots$ | $\frac{m+2}{4} \cdot \frac{8}{3m+6} = \frac{8(m+2)}{4(3m+6)} = \ldots$

Rule 2: $\frac{a}{b} \cdot c = \frac{a}{b} \cdot \frac{c}{1}$ and $\frac{a}{b} \div c = \frac{a}{b} \div \frac{c}{1}$

$\frac{3}{4} \cdot 16 = \frac{3}{4} \cdot \frac{16}{1} = \ldots$ | $\frac{6}{2x+10} \cdot (x^2+3x-10) = \frac{6}{2x+10} \cdot \frac{x^2+3x-10}{1} = \ldots$

Rule 3: $\frac{a}{b} \div \frac{c}{d} = \frac{a}{b} \cdot \frac{d}{c}$

$\frac{1}{12} \div \frac{2}{3} = \frac{1}{12} \cdot \frac{3}{2} = \ldots$ | $\frac{3ab^4}{a^3} \div \frac{3ab}{ab^2} = \frac{3ab^4}{a^3} \cdot \frac{ab^2}{3ab} = \ldots$

**Finish multiplying and dividing the rational expressions given above so that each is simplified completely.**

1. Rule 1

$\frac{2}{3}$

2. Rule 2

$3(x-2)$

3. Rule 3

$\frac{b^5}{a^2}$

**Multiply or divide. Simplify your answer.**

4. $\frac{3}{t+2} \cdot (t^2 - 4)$

$3t - 6$

5. $\frac{m+5}{4} \div (3m + 15)$

$\frac{1}{12}$

6. $\frac{x^2y^3}{8y} \cdot \frac{10x^4}{xy^2}$

$\frac{5x^5}{4}$

LESSON 12-5

## Practice A

### Adding and Subtracting Rational Expressions

**Add or subtract. Simplify your answer.**

1. $\frac{6x}{5x^2} + \frac{4x}{5x^2}$ — $\frac{2}{x}$

2. $\frac{t}{t^2-36} + \frac{6}{t^2-36}$ — $\frac{1}{t-6}$

3. $\frac{c+5}{3c+3} - \frac{4}{3c+3}$ — $\frac{1}{3}$

4. $\frac{3b}{b^2+8b+15} - \frac{b-10}{b^2+8b+15}$ — $\frac{2}{b+3}$

**Find the LCM of the given expressions.**

5. $3x^2y^2, 9x^3y$ — $9x^3y^2$

6. $a^2-16, (a-4)(a+5)$ — $(a+4)(a-4)(a+5)$

**Use a common denominator to add or subtract. Simplify your answer.**

7. $\frac{7}{3m} - \frac{2}{m}$ — $\frac{1}{3m}$

8. $\frac{x^2+x}{x^2+6x+5} + \frac{4}{x+5}$ — $\frac{x+4}{x+5}$

9. $\frac{3}{x-10} + \frac{9}{10-x}$ — $\frac{-6}{x-10}$

10. $\frac{3}{x} - \frac{4}{x^2-9x}$ — $\frac{3x-31}{x^2-9x}$

11. Carlie biked 6 miles on mountain trails and 15 miles on the road. Her biking rate on the road was 3 times faster than that on the trails. Let $r$ represent her rate on the trails in mi/h.

   a. Write and simplify an expression that represents the number of hours she biked in all. $\left(Hint: t = \frac{d}{r}\right)$ — $\frac{11}{r}$

   b. Find the time if her rate on the trails was 3 mi/h. — $3\frac{2}{3}$ h or 3 h 40 min




LESSON 12-5

## Practice B

### Adding and Subtracting Rational Expressions

**Add or subtract. Simplify your answer.**

1. $\frac{3m}{8m^3} + \frac{m}{8m^3}$ — $\frac{1}{2m^2}$

2. $\frac{x^2-6x}{x+3} + \frac{4x-15}{x+3}$ — $x-5$

3. $\frac{c^2+c}{c^2-25} - \frac{c^2+5}{c^2-25}$ — $\frac{1}{c+5}$

4. $\frac{6a-1}{a^2+7a+10} - \frac{2a-9}{a^2+7a+10}$ — $\frac{4}{a+5}$

**Find the LCM of the given expressions.**

5. $4a^2b, 6a, 10b^3$ — $60a^2b^3$

6. $x^2+5x+6, (x+3)(x-1)$ — $(x+2)(x+3)(x-1)$

**Add or subtract. Simplify your answer.**

7. $\frac{5}{3n} - \frac{2}{2n}$ — $\frac{2}{3n}$

8. $\frac{y^2+4y}{y^2+6y+8} + \frac{3}{y+2}$ — $\frac{y+3}{y+2}$

9. $\frac{x+2}{x^2-9} - \frac{1}{9-x^2}$ — $\frac{1}{x-3}$

10. $\frac{1}{6y^2+24y} - \frac{3}{y^2-y-20}$ — $\frac{-17y-5}{6y^3-6y^2-120y}$

11. Kendrick walked 1 mile, and then jogged 5 miles. His jogging speed was 4 times his walking speed $w$ in mi/h.

   a. Write and simplify an expression that represents Kendrick's total exercise time. — $\frac{9}{4w}$

   b. How many minutes did Kendrick exercise if his walking speed was 3 mi/h? — 45 min




LESSON 12-5

## Practice C

### Adding and Subtracting Rational Expressions

**Add or subtract. Simplify your answer.**

1. $\frac{7y}{2y^3} + \frac{3y}{2y^3}$ — $\frac{5}{y^2}$

2. $\frac{8t+5}{t^2+t-12} - \frac{5t-7}{t^2+t-12}$ — $\frac{3}{t-3}$

3. $\frac{7m+8}{4m^2-25} - \frac{m-7}{4m^2-25}$ — $\frac{3}{2m-5}$

4. $\frac{3}{p} + \frac{7}{4p}$ — $\frac{19}{4p}$

5. $\frac{x^2+2x}{x^2+5x+6} - \frac{5}{x+3}$ — $\frac{x-5}{x+3}$

6. $\frac{4k}{k-6} + \frac{k}{6-k}$ — $\frac{3k}{k-6}$

7. $\frac{x}{2x-10} + \frac{1}{3x-15}$ — $\frac{3x+2}{6x+30}$

8. $\frac{c}{c^2-d^2} + \frac{d}{c^2}$ — $\frac{c^3+c^2d-d^3}{c^4-c^2d^2}$

9. $\frac{x}{x^2-4x-5} + \frac{1}{x+1} + \frac{1}{5}$ — $\frac{x^2+6x-30}{5x^2-20x-25}$

10. $\frac{2}{t-1} + \frac{t}{t^2-2t+1} - \frac{3}{t+1}$ — $\frac{7t-5}{t^3-t^2-t+1}$

11. A boat sailed from a dock 35 miles up a river and back. The boat's average sailing rate is 6 times the speed of the river's current.

   a. Write and simplify an expression that represents the time it took the boat to make the round trip. — $\frac{12}{x}$

   b. The boat left the dock at 3:15 P.M. What time will the boat return if the rate of the river is 4.5 mi/h? — 5:55 P.M.




LESSON 12-5

## Reteach

### Adding and Subtracting Rational Expressions

The rules for adding and subtracting rational expressions are the same as the rules for adding and subtracting fractions.

$$\frac{2}{7} + \frac{4}{7} = \frac{2+4}{7} = \frac{6}{7} \qquad \frac{4}{5} - \frac{1}{5} = \frac{4-1}{5} = \frac{3}{5}$$

**Add** $\frac{5x+10}{x^2-16} + \frac{10}{x^2-16}$.

**Simplify your answer.**

$\frac{5x+10}{x^2-16} + \frac{10}{x^2-16}$

$\frac{(5x+10)+10}{x^2-16}$ *Add numerators.*

$\frac{5x+20}{x^2-16}$ *Add like terms.*

$\frac{5(x+4)}{(x+4)(x-4)}$ *Factor.*

$\frac{5(x+4)^1}{(x+4)^1(x-4)}$ *Simplify.*

$\frac{5}{x-4}$

**Subtract** $\frac{2x}{2x^2+6} - \frac{x-3}{2x^2+6}$.

**Simplify your answer.**

$\frac{2x}{2x^2+6} - \frac{x-3}{2x^2+6}$

$\frac{2x-(x-3)}{2x^2+6}$ *Subtract numerators.*

$\frac{2x-x+3}{2x^2+6}$ *Distribute −1.*

$\frac{x+3}{2x^2+6}$ *Combine like terms.*

$\frac{x+3}{2(x+3)}$ *Factor.*

$\frac{(x+3)^1}{2(x+3)^1}$ *Simplify.*

$\frac{1}{2}$

**Add or subtract. Simplify your answer.**

1. $\frac{x+2}{x^2-100} + \frac{8}{x^2-100}$ — $\frac{1}{x-10}$

2. $\frac{x^2-18}{x+6} + \frac{3x}{x+6}$ — $x-3$

3. $\frac{x}{5x+30} + \frac{6}{5x+30}$ — $\frac{1}{5}$

4. $\frac{x}{10x^2-20x} - \frac{2}{10x^2-20x}$ — $\frac{1}{10x}$

5. $\frac{x^2}{x^2+10x+25} - \frac{25}{x^2+10x+25}$ — $\frac{x-5}{x+5}$

6. $\frac{x^2}{x+6} - \frac{12-4x}{x+6}$ — $x-2$

LESSON 12-5

## Reteach

### Adding and Subtracting Rational Expressions (continued)

Like fractions, you may need to multiply by a form of 1 to obtain common denominators.

$$\frac{1}{6}+\frac{4}{15}=\frac{1}{6}\left(\frac{5}{5}\right)+\frac{4}{15}\left(\frac{2}{2}\right)=\frac{5}{30}+\frac{8}{30}=\frac{13}{30}$$

**Add $\frac{3}{10x}+\frac{2}{5x^2}$. Simplify your answer.**

$\frac{3}{10x}+\frac{2}{5x^2}$ *The LCD is $10x^2$.*

$\frac{3}{10x}\left(\frac{x}{x}\right)+\frac{2}{5x^2}\left(\frac{2}{2}\right)$ *Multiply each term by a form of 1.*

$\frac{3x}{10x^2}+\frac{4}{10x^2}$ *Multiply.*

$\frac{3x+4}{10x^2}$ *Add numerators.*

**Subtract $\frac{3}{x-4}-\frac{5}{4-x}$. Simplify your answer.**

$\frac{3}{x-4}-\frac{5}{4-x}$ *The LCD is $x-4$.*

$\frac{3}{x-4}-\frac{5}{4-x}\left(\frac{-1}{-1}\right)$ *Multiply the second term by a form of 1.*

$\frac{3}{x-4}-\frac{-5}{x-4}$ *Multiply.*

$\frac{8}{x-4}$ *Add numerators.*

7. Identify the LCD of $\frac{5}{8x^2}$ and $\frac{13}{20x}$. $40x^2$

8. Identify the LCD of $\frac{3}{5x}$ and $\frac{x}{x+7}$. $5x(x+7)$

**Add or subtract. Simplify your answer.**

9. $\frac{2x}{8x^3}+\frac{5}{12x}$

$\frac{2x}{8x^3}\left(\frac{3}{3}\right)+\frac{5}{12x}\left(\frac{2x^2}{2x^2}\right)$

$\frac{5x+3}{12x^2}$

10. $\frac{5x+11}{x^2+5x+6}-\frac{4}{x+3}$

$\frac{5x+11}{(x+3)(x+2)}-\frac{4}{x+3}\left(\frac{x+2}{x+2}\right)$

$\frac{1}{x+2}$

11. $\frac{4x}{6x}+\frac{2}{x+4}$ $\frac{2x+14}{3x+12}$

12. $\frac{3x-6}{x^2-36}-\frac{2}{x+6}$ $\frac{1}{x-6}$

13. $\frac{2x}{x^2+4x+3}+\frac{1}{x+1}$ $\frac{x}{x+3}$



LESSON 12-5

## Challenge

### Average Speed

**At 8 A.M., Mr. Varhalla drives from home to work. Because of rush-hour traffic, he is only able to drive 40 mi/h.**

**At 3 P.M. he leaves work early to drive back home. Because he avoids rush-hour traffic, he is able to drive the same distance at 65 mi/h.**

1. Let $d$ represent the distance in miles between home and work. What is the total distance of Mr. Varhalla's round trip? $2d$
2. Write an expression for the time in hours that it takes Mr. Varhalla to drive from home to work. $\frac{d}{40}$
3. Write an expression for the time in hours that it takes him to drive from work back home. $\frac{d}{65}$
4. Write (but don't simplify) an addition expression for the total time in hours of Mr. Varhalla's round trip. $\frac{d}{40}+\frac{d}{65}$
5. Because $d = rt$, you can find average speed for the entire round trip by dividing the total distance by the total time, or $r=\frac{d}{t}$. Write a complex fraction that represents Mr. Varhalla's average speed in miles per hour. $\frac{2d}{\frac{d}{40}+\frac{d}{65}}$
6. Simplify the complex fraction to find Mr. Varhalla's average speed in miles per hour. Round to the nearest tenth. ≈49.5 mi/h
7. Is the average speed for the round trip the same as the average of the speeds 40 mi/h and 65 mi/h?

No; the average of 40 mi/h and 65 mi/h is $\frac{40+65}{2}=52.5$ mi/h.

**Use what you have learned in questions 1–7 to solve these problems. For each, first write a complex fraction that represents average speed and then simplify it to find the average speed. Round your answers to the nearest tenth.**

8. Teri drives 30 mi/h from home to the hardware store. After loading her truck with lumber, she drives 15 mi/h back home. What is Teri's average speed in miles per hour for the entire round trip? 20 mi/h
9. Kevin runs four laps around Piedmont Park at 10 mi/h. Then, to cool down, he walks one more lap at 3 mi/h. What is Kevin's average speed for the entire workout? (*Hint:* Let $d$ represent the distance of one lap.) 6.8 mi/h



LESSON 12-5

## Problem Solving

### Adding and Subtracting Rational Expressions

**Adib is kayaking on the Peconic River in Long Island, New York. He paddles his kayak at an average rate of 3 mi/h, but does not know the rate of the river's current. Adib plans to kayak upstream 2 mi and then back downstream to his starting point.**

1. Let $x$ represent the rate of the current in the Peconic River in miles per hour. Write and simplify an expression for the total time of Adib's round trip. $\frac{12}{(3-x)(3+x)}$
2. If the rate of the river's current is 2 mi/h, how long will it take Adib to kayak round trip? $\frac{12}{5}$ h, or 2.4 h, or 2 h 24 min
3. If the rate of the river's current is 3 mi/h, how long will it take Adib to kayak round trip? Explain what your answer could mean in this context.

$\frac{12}{0}$ = undefined; because the rate of the current is equal to Adib's rate of paddling, he remains stationary and is never able to kayak upstream.

**Select the best answer.**

4. Terry drives 2 mi on city streets, and 20 mi on the highway. Her speed on the highway is three times her speed on the city streets $r$, in miles per hour. Write and simplify and expression that represents the length of Terry's trip in hours.

A $\frac{13}{3r}$ Ⓒ $\frac{26}{3r}$

B $\frac{22}{4r}$ D $\frac{62}{3r}$

5. Nahuel walks 1 km to school at a rate of $w$ km/h. When he gets to school, he realizes that he forgot his math book, so he jogs back home and back to school at a rate 4 km/h faster than his walking rate. Write and simplify an expression that represents the length of Nahuel's entire trip to school in hours.

F $\frac{3}{2w}$ H $\frac{2(w+1)}{w(w+2)}$

G $\frac{3}{2(w+2)}$ Ⓙ $\frac{3w+4}{w(w+4)}$

6. A test consists of 4 free response questions and 50 multiple choice questions. The test's writers assume an average student can do $x$ free response questions per hour, and four times as many multiple choice questions per hour. If an average student can do 6 free response questions per hour, how long is the test?

A $1\frac{4}{5}$ h C $3\frac{3}{8}$ h

Ⓑ $2\frac{3}{4}$ h D $5\frac{2}{3}$ h

7. In a carnival game, Beth runs 100 m to a table, picks up water balloons, and runs back to the starting line. Beth originally runs to the table at a rate of $y$ m/s, but runs 1 m/s slower when she carries the balloons. If Beth originally runs 5 m/s, how long, to the nearest second, does it take her to finish?

F 10 s H 37 s

G 22 s Ⓙ 45 s



LESSON 12-5

## Reading Strategies

### Follow a Procedure

When adding or subtracting rational expressions, there may be several steps that need to be taken. Look at the example below.

**Add $\frac{3}{9x}+\frac{x}{6x^3}$.**

$9x = 3\cdot 3\cdot x$

$6x^3 = 2\cdot 3\cdot x\cdot x\cdot x$

LCD: $2\cdot 3\cdot 3\cdot x\cdot x\cdot x = 18x^3$ — 1 Identify the least common denominator, LCD.

$\frac{3}{9x}\left(\frac{2x^2}{2x^2}\right)+\frac{x}{6x^3}\left(\frac{3}{3}\right)$ — 2 Multiply expressions by appropriate form of 1.

$\frac{6x^2}{18x^3}+\frac{3x}{18x^3}$ — 3 Write each expression with the new LCD.

$\frac{6x^2+3x}{18x^3}$ — 4 Add/subtract numerators. Keep denominator.

$\frac{3x(2x+1)}{18x^3}$ — 5 Factor the numerator and/or denominator.

$\frac{2x+1}{6x^2}$ — 6 Simplify as needed.

**Answer each question about the procedure shown above.**

1. If the two expressions have the same denominator, which step can you start with? Step 4
2. Why was $\left(\frac{2x^2}{2x^2}\right)$ the "appropriate form of 1" for the first expression?

When multiplied with $9x$, it gave the LCD $18x^3$.

3. How would the answer to this problem change if it were subtraction instead of addition?

The numerator would be $2x-1$.

**Add or subtract.**

4. $\frac{x+5}{x^2-9}-\frac{2}{x^2-9}$ $\frac{1}{x-3}$

5. $\frac{1}{4x}+\frac{4x}{6x^2}$ $\frac{11}{12x}$

## LESSON 12-6 Practice A
### Dividing Polynomials

**Divide each term of the polynomial by the monomial.**

1. $(12m^3 + 4m^2) \div 4m$

$\frac{12m^3}{4m} + \frac{4m^2}{4m}$

**$3m^2 + m$**

2. $(40y^5 + 24y^3 - 20y^2) \div 4y^2$

**$10y^3 + 6y - 5$**

3. $(15x^5 + 15x^2 - 10x) \div 5x^2$

**$3x^3 + 3 - \frac{2}{x}$**

4. $(6w^7 - 2w^3 - 7) \div 2w^4$

**$3w^3 - \frac{1}{w} - \frac{7}{2w^4}$**

**Divide each polynomial by the binomial.**

5. $(k^2 + 14k + 45) \div (k + 9)$

**$k + 5$**

6. $(x^2 - 5x - 14) \div (x + 2)$

**$x - 7$**

**Divide using long division. There will not be a remainder.**

7. $a + 3\overline{)a^2 + 4a + 3}$

**$a + 1$**

8. $t + 4\overline{)2t^2 + 11t + 12}$

**$2t + 3$**

**Divide using long division. There will be a remainder.**

9. $(c^2 + 7c + 14) \div (c + 4)$

**$c + 3 + \frac{2}{c + 4}$**

10. $(k^2 + 5) \div (k - 2)$
(*Hint*: Rewrite $k^2 + 5$ as $k^2 + 0k + 5$.)

**$k + 2 + \frac{9}{k - 2}$**




## LESSON 12-6 Practice B
### Dividing Polynomials

**Divide.**

1. $(15c^3 + 3c^2) \div 3c$

**$5c^2 + c$**

2. $(20b^4 - 12b + 4) \div 4b$

**$5b^3 - 3 + \frac{1}{b}$**

3. $(27q^6 - 3q^3 + 18) \div 9q^5$

**$3q - \frac{1}{3q^2} + \frac{2}{q^5}$**

4. $(15t^4 - 30t^2 + 6) \div 15t^3$

**$t - \frac{2}{t} + \frac{2}{5t^3}$**

5. $(d^2 - 4d - 77) \div (d - 11)$

**$d + 7$**

6. $(x^2 - 12x + 27) \div (x - 3)$

**$x - 9$**

7 $(9p^2 + 6p + 1) \div (3p + 1)$

**$3p + 1$**

8. $(4b^2 + b - 3) \div (b + 1)$

**$4b - 3$**

**Divide using long division.**

9. $\frac{m^2 + 4m - 12}{m + 6}$

**$m - 2$**

10. $(12y^2 + 31y + 14) \div (y + 2)$

**$12y + 7$**

11. $(t^2 + t - 6) \div (t - 1)$

**$t + 2 + \frac{-4}{t - 1}$**

12. $(3p^3 + 4p - 6) \div (p + 2)$

**$3p^2 - 6p + 16 + \frac{-38}{p + 2}$**




## LESSON 12-6 Practice C
### Dividing Polynomials

**Divide.**

1. $(3x^2 + x) \div 3x$

**$x + \frac{1}{3}$**

2. $(12x^2 - 7x + 16) \div -2x^2$

**$-6 + \frac{7}{2x} - \frac{8}{x^2}$**

3. $(k^6 - 2k^5 + 10k^3 - 20) \div 5k^5$

**$\frac{k}{5} - \frac{2}{5} + \frac{2}{k^2} - \frac{4}{k^5}$**

4. $(-3m^5n^4 + 12m^3n^2 - mn) \div -6m^2n^2$

**$\frac{m^3n^2}{2} - 2m + \frac{1}{6mn}$**

5. $(4t^2 + 17t - 42) \div (t + 6)$

**$4t - 7$**

6. $(24x^2 + x - 10) \div (8x - 5)$

**$3x + 2$**

7. $(48b^2 + 32b + 5) \div (4b + 1)$

**$12b + 5$**

8. $(3y^3 + 15y^2 + y + 5) \div (y + 5)$

**$3y^2 + 1$**

**Divide using long division.**

9. $\frac{2n^2 - 11n + 12}{2n - 3}$

**$n - 4$**

10. $(3a^2 + 7a - 20) \div (a + 4)$

**$3a - 5$**

11. $(2y^2 + 3y - 5) \div (y - 3)$

**$2y + 9 + \frac{22}{y - 3}$**

12. $(4t^2 + 3t^3 - 8) \div (6 + t)$

**$3t^2 - 14t + 84 + \frac{-512}{t + 6}$**




## LESSON 12-6 Reteach
### Dividing Polynomials

To divide a polynomial by a monomial, first write the division as a rational expression.

**Divide $(12x^2 + 9x) \div 3x$.**

$\frac{12x^2 + 9x}{3x}$ *Rewrite as a rational expression.*

$\frac{12x^2}{3x} + \frac{9x}{3x}$ *Divide each term by 3x.*

$\frac{\cancel{12}^4 x^{\cancel{2}^1}}{_1\cancel{3}\cancel{x}_1} + \frac{\cancel{9}^3 \cancel{x}^1}{_1\cancel{3}\cancel{x}_1}$ *Divide out common factors.*

$4x + 3$ *Simplify.*

To divide a polynomial by a binomial, try to factor and divide out common factors.

**Divide $\frac{x^2 + 6x + 5}{x + 1}$.**

$\frac{x^2 + 6x + 5}{x + 1}$

$\frac{(x + 1)(x + 5)}{x + 1}$ *Factor the numerator.*

$\frac{\cancel{(x + 1)}^1(x + 5)}{\cancel{x + 1}_1}$ *Divide out common factors.*

$x + 5$ *Simplify.*

You may find that a rational expression does not divide evenly.

**Divide $(5x^2 + 10x + 3) \div 5x$.**

$\frac{5x^2 + 10x + 3}{5x}$ *Rewrite as a rational expression.*

$\frac{5x^2}{5x} + \frac{10x}{5x} + \frac{3}{5x}$ *Divide each term by 5x.*

$x + 2 + \frac{3}{5x}$

**Divide.**

1. $(14x^3 + 6x) \div 2x$

**$7x^2 + 3$**

2. $(6x^4 + 12x^2 + 18x) \div 6x^2$

**$x^2 + 2 + \frac{3}{x}$**

3. $(10x^3 + 12x^2 + 2x) \div 2x$

**$5x^2 + 6x + 1$**

4. $(x^2 - 49) \div (x + 7)$

**$x - 7$**

5. $(3x^2 - 15x) \div (x - 5)$

**$3x$**

6. $(x^2 - 5x - 24) \div (x - 8)$

**$x + 3$**

**LESSON 12-6** 

# Reteach

## Dividing Polynomials (continued)

You can use long division to divide a polynomial by a binomial.

**Divide $(2x^2 + 2x + 7) \div (x + 3)$.**

$x + 3\overline{)2x^2 + 2x + 7}$ *Write in long division form.*

$2x - 4$ *Think: $x \cdot ? = 2x^2$. Use $2x$.*

$x + 3\overline{)2x^2 + 2x + 7}$

$-(2x^2 + 6x)$ *Think: $x \cdot ? = -4x$. Use $-4$.*

$-4x + 7$

$-(-4x - 12)$

$19$

$(2x^2 + 2x + 7) \div (x + 3) = 2x - 4 + \frac{19}{x + 3}$

Use a zero coefficient if the polynomial is missing a term.

**Divide $(x^2 - 4) \div (x + 2)$.**

$x + 2\overline{)x^2 + 0x - 4}$ *Write in long division form.*

$x - 2$

$x + 2\overline{)x^2 + 0x - 4}$ *Think: $x \cdot ? = x^2$. Use $x$.*

$-(x^2 + 2x)$

$-2x - 4$ *Think: $x \cdot ? = -2x$. Use $-2$.*

$-(-2x - 4)$

$0$

$(x^2 - 4) \div (x + 2) = (x - 2)$

**Divide using long division.**

7. $(x^2 - 4x - 12) \div (x + 2)$

$\boxed{x} - \boxed{6}$

$x + 2\overline{)x^2 + 4x - 12}$

$-(\boxed{x^2 + 2x})$

$\boxed{-6x} - \boxed{12}$

$-(\boxed{-6x - 12})$

$0$

$x - 6$

8. $(x^2 + 6x + 3) \div (x + 4)$

$\boxed{x} + \boxed{2}$

$x + 4\overline{)x^2 + 6x + 3}$

$-(\boxed{x^2 + 4x})$

$\boxed{2x} + \boxed{3}$

$-(\boxed{2x + 8})$

$-5$

$x + 2 + \frac{-5}{x + 4}$

9. $(x^2 - 25) \div (x + 5)$

$x - 5$

10. $(x^2 + 5x + 4) \div (x + 2)$

$x + 3 + \frac{-2}{x + 2}$




**LESSON 12-6**

# Challenge

## The Remainder Theorem

**For 1–5, use the polynomial $P(x) = 2x^2 + 3x - 5$.**

1. Use long division to divide $P(x) \div (x - 2)$.
   a. What value do you get for the remainder? **9**
   b. Now evaluate $P(2)$. What value do you get? **9**
2. Use long division to divide $P(x) \div (x - 1)$.
   a. What value do you get for the remainder? **0**
   b. Now evaluate $P(1)$. What value do you get? **0**
3. Use long division to divide $P(x) \div x$.
   a. What value do you get for the remainder? **−5**
   b. Now evaluate $P(0)$. What value do you get? **−5**
4. Use long division to divide $P(x) \div (x + 1)$.
   a. What value do you get for the remainder? **−6**
   b. Now evaluate $P(-1)$. What value do you get? **−6**
5. Use long division to divide $P(x) \div (x + 2)$.
   a. What value do you get for the remainder? **−3**
   b. Now evaluate $P(-2)$. What value do you get? **−3**
6. Look for a pattern in your results and complete this theorem.
   **Polynomial Remainder Theorem:** If a polynomial $P(x)$ is divided by $(x - r)$, then the remainder is a constant given by **$P(r)$**.
7. Your results in question 2 are unique. What do they tell you about the binomial $(x - 1)$?
   **Because the remainder is zero, $(x - 1)$ is a factor of $2x^2 + 3x - 5$: $(x - 1)(2x + 5)$.**
8. Given $Q(x) = 6x^4 - 31x^3 - 49x^2 + 104x + 60$, explain how you could determine whether a binomial of the form $(x - r)$ is a factor of $Q(x)$ without doing long division.
   **Evaluate $Q(r)$; if $Q(r) = 0$, then the binomial $(x - r)$ is a factor.**
9. a. Is $(x - 1)$ a factor of $Q(x)$? Justify.
   **No; $Q(1) = 6(1)^4 - 31(1)^3 - 49(1)^2 + 104(1) + 60 = 90 \neq 0$**
   b. Is $(x + 2)$ a factor of $Q(x)$? Justify.
   **Yes; $Q(-2) = 6(-2)^4 - 31(-2)^3 - 49(-2)^2 + 104(-2) + 60 = 0$**




**LESSON 12-6**

# Problem Solving

## Dividing Polynomials

**Write the correct answer.**

1. The area of a rectangle is $x^2 + 2x - 8\text{m}^2$, and the width is $x - 2$ m. Find the length.
   **$x + 4$ m**
2. On the dartboard below, the area of the outside ring is the difference between the area of the two largest circles, or $\pi(x + 4)^2 - \pi(x + 2)^2$ in². The area of the bull's-eye is $\pi x^2$ in². Expand and divide $[\pi(x + 4)^2 - \pi(x + 2)^2] \div \pi x^2$ to find the ratio of the area of the outside ring to the bull's-eye.

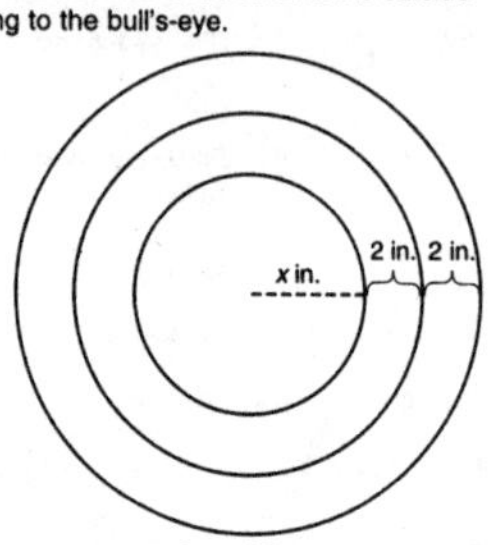


   **$\frac{4}{x} + \frac{12}{x^2}$**
3. The area of a rectangular piece of paper is $4n^2 - 4n - 9$ cm². Then a triangle with area $2n + 6$ cm² is cut from one corner of the paper. What is the ratio of the area of the paper that remains to the area of the triangle that was removed? (*Hint:* First subtract to find the area of the paper that remains.)
   **$2n - 9 + \frac{39}{2n + 6}$**

**Select the best answer.**

4. The area of a triangle is $8a^2 - 10a$ ft², and the height is $4a$ ft. Find the base. (*Hint:* Solve $A = \frac{1}{2}bh$ for $b$.)
   A $a - \frac{5a}{4}$ ft
   B $2a - \frac{5a}{2}$ ft
   C $4a - \frac{5a}{2}$ ft
   **(D) $4a - 5$ ft**
5. The volume of a rectangular prism is $2x^3 + 9x^2 - 11x - 30$ cm³, and the height is $x + 5$ cm. Find the area of the base. ($V = BH$, where $B$ is the area of the base and $H$ is the height.)
   **(F) $2x^2 - x - 6$ cm²**
   G $2x^2 - x - 16 + \frac{-110}{x + 5}$ cm²
   H $2x^2 + 4x - 16 + \frac{-35}{x + 5}$ cm²
   J $2x^2 + 19x + 84 + \frac{390}{x + 5}$ cm²
6. The volume of a cylinder is $V = \pi r^2 h$, where $r$ is the radius and $h$ is the height. The volume of a certain cylinder is $\pi(n^3 - 6n^2 - 36n + 216)$ in³, and the radius is $n - 6$ in. Find the height. (*Hint:* To divide by $r^2$, divide by $r$ twice.)
   A $n$ in.
   B $n - 6$ in.
   **(C) $n + 6$ in.**
   D $n^2 - 36$ in.




**LESSON 12-6**

# Reading Strategies

## Use a Model

The process for dividing polynomials is the same as the process for dividing whole numbers. The steps are shown in the model below.

**Divide $(x^2 + 5x - 4) \div (x + 2)$.**

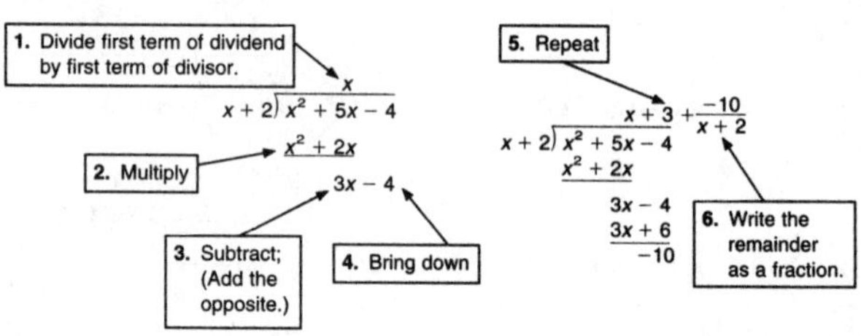


The quotient is $x + 3 + \frac{-10}{x + 2}$.

**Use the model above to answer the following.**

1. What division took place in step 1? **$x^2 \div x = x$**
2. In step 2, what was multiplied? **$x(x + 2)$**
3. In step 3, polynomials are subtracted by "adding the opposite." What is the opposite of $x^2 + 2x$? **$-x^2 - 2x$**
4. What term was brought down in step 4? **$-4$**
5. In step 5, what is meant by "repeat"? **Divide, multiply, and subtract with $3x - 4$ as the dividend.**

**Divide using long division.**

6. $(x^2 + 8x + 15) \div (x + 3)$
   **$x + 5$**
7. $(x^2 - 2x + 7) \div (x + 3)$
   **$x - 5 + \frac{22}{x + 3}$**

LESSON 12-7

## Practice A

### Solving Rational Equations

**Solve using cross products. Check your answer.**

1. $\frac{6}{b+4} = \frac{4}{b}$ **8**

2. $\frac{2}{x} = \frac{3}{x-5}$ **−10**

**Solve using the LCD. Check your answer.**

3. $\frac{9}{c} - \frac{8}{c} = \frac{1}{4}$ **4**

4. $\frac{2}{n+1} + \frac{3}{n} = \frac{9}{2n}$ **3**

5. $\frac{9}{2x} + \frac{7}{x} = \frac{12x-1}{2x}$ **2**

6. $\frac{3}{x-2} + \frac{3}{x+2} = \frac{9}{x+2}$ **6**

**Solve using cross products. Identify any extraneous solutions.**

7. $\frac{2}{x-4} = \frac{x-1}{x-4}$ **3; 4 is an extraneous solution**

8. $\frac{x-2}{x+2} = \frac{3}{x}$ **−1 and 6; no extraneous solutions**

9. Marlena can rake the leaves in 4 hours. Joshua can rake the same amount of leaves in 2 hours. How long will it take them to rake the leaves if they work together? **$1\frac{1}{3}$h or 1h 20min**

**One number is 4 times another number. What are the numbers if the sum of their reciprocals is $\frac{1}{4}$?**

10. If one of the numbers is $x$, what is the other number? **4x**

11. Write the reciprocal of each number. **$\frac{1}{x}, \frac{1}{4x}$**

12. Write and solve an equation showing that the sum of the reciprocals is $\frac{1}{4}$. **$\frac{1}{x} + \frac{1}{4x} = \frac{1}{4}$**

**$x = 5$**

13. What are the two numbers? **5 and 20**




LESSON 12-7

## Practice B

### Solving Rational Equations

**Solve. Check your answer.**

1. $\frac{6}{t+3} = \frac{4}{t}$ **6**

2. $\frac{3}{m} = \frac{4}{m-2}$ **−6**

3. $\frac{a}{4} + \frac{1}{2} = \frac{2}{3}$ **$\frac{2}{3}$**

4. $\frac{3}{2x} - \frac{3}{x-2} = \frac{1}{2x}$ **−1**

5. $\frac{3}{2x} + \frac{5}{x} = \frac{13}{x+4}$ **4**

6. $\frac{3}{x} + \frac{3x+1}{x^2} = \frac{13}{x^2}$ **2**

**Solve. Identify any extraneous solutions.**

7. $\frac{8}{x-2} = \frac{x+3}{x-2}$ **5; 2 is an extraneous solution.**

8. $\frac{-2}{x-1} = \frac{x-8}{x+1}$ **5 and 2; no extraneous solutions**

9. Caroline can paint a fence in 6 hours. Her sister Lily can paint the same fence in 4 hours. How long will it take them to paint the fence if they work together? **$2\frac{2}{5}$h or 2h 24 min**

10. Jalon bicycled against the wind for 10 miles in the same time he bicycled with the wind for 25 miles. The wind speed was 4 mi/h. What was Jalon's average bicycling speed? (*Hint*: Use $t = \frac{d}{r}$.) **$9\frac{1}{3}$mi/h**

11. There are two positive numbers. The second number is 6 less than the first number. When the reciprocal of the second number is subtracted from the reciprocal of the first, the difference is $-\frac{3}{8}$. Find the first number. **8**




LESSON 12-7

## Practice C

### Solving Rational Equations

**Solve. Check your answer.**

1. $\frac{4}{y} = \frac{1}{y+6}$ **−8**

2. $\frac{5}{3x+1} = 2$ **$\frac{1}{2}$**

3. $\frac{4}{c+2} + \frac{5}{c} = \frac{21}{3c}$ **2**

4. $\frac{2}{n} = \frac{15}{n^2} - 1$ **−5, 3**

5. $\frac{5}{6} - \frac{1}{x} = \frac{3x+8}{6x}$ **7**

6. $\frac{4}{x+3} + \frac{3}{x} = \frac{4x+1}{5x}$ **$-\frac{3}{2}$, 7**

7. $\frac{3}{x+3} = \frac{x-2}{x+3}$ **5**

8. $\frac{x}{x^2-8} = \frac{2}{x}$ **−4 and 4**

9. Trey can clean a room in 40 minutes. His brother Marcus can clean the same room in 20 minutes. How long will it take them to clean the room if they work together? **$13\frac{1}{3}$min**

10. A boat traveled against the current for 70 miles for the same time it traveled with the current for 95 miles. What was the boat's average speed if the rate of the current was 5 mi/h? **33 mi/h**

11. The reciprocal of a number minus the reciprocal of two less than the number is $-\frac{2}{15}$. Find the number. **−3 or 5**

12. Pump A can fill a tank 3 hours faster than Pump B. If both pumps work together, it takes 8 hours to fill the tank. Working alone, how long would it take Pump A to fill the tank? Round to the nearest tenth of an hour. (*Hints*: Use the Quadratic Formula and make sure times are positive for both pumps.) **14.6 h**




LESSON 12-7

## Reteach

### Solving Rational Equations

A **rational equation** is an equation that contains one or more rational expressions. Some rational equations are proportions and can be solved using cross products. Solutions to all rational equations must be checked.

**Solve $\frac{4}{x-3} = \frac{2}{x}$.**

$\frac{4}{x-3} \times \frac{2}{x}$

$4(x) = 2(x-3)$ *Multiply.*

$4x = 2x - 6$ *Distribute.*

$-2x \quad -2x$ *Add −2x to both sides.*

$2x = -6$

$\frac{2x}{2} = \frac{-6}{2}$ *Divide.*

$x = -3$ *Simplify.*

***Check:***

| $\frac{4}{x-3}$ | $= \frac{2}{x}$ |
|---|---|
| $\frac{4}{(-3)-3}$ | $\frac{2}{(-3)}$ |
| $\frac{4}{-6}$ | $\frac{2}{-3}$ |
| $-\frac{2}{3}$ | $-\frac{2}{3}$ ✓ |

The solution is −3.

**Solve $\frac{x-4}{x^2-4} = \frac{-2}{x-2}$.**

$\frac{x-4}{x^2-4} \times \frac{-2}{x-2}$

$(x-4)(x-2) = -2(x^2-4)$

$x^2 - 6x + 8 = -2x^2 + 8$

$+2x^2 \qquad +2x^2$

$3x^2 - 6x + 8 = 8$

$-8 \quad -8$

$3x^2 - 6x = 0$

$3x(x-2) = 0$ *Zero Product Property*

$x = 0$ or $x = 2$

***Check:***

| $\frac{x-4}{x^2-4}$ | $= \frac{-2}{x-2}$ |
|---|---|
| $\frac{(0)-4}{(0)^2-4}$ | $\frac{-2}{(0)-2}$ |
| $\frac{-4}{-4}$ | $\frac{-2}{-2}$ |
| 1 | 1 ✓ |

| $\frac{x-4}{x^2-4}$ | $= \frac{-2}{x-2}$ |
|---|---|
| $\frac{(2)-4}{(2)^2-4}$ | $\frac{-2}{(2)-2}$ |
| $\frac{-2}{0}$ | $\frac{-2}{0}$ ✗ |
| undefined | |

The only solution is 0.

**Solve. Check your answer.**

1. $\frac{3}{x+2} = \frac{4}{x+1}$ **$x = -5$**

2. $\frac{x}{6} = \frac{x}{x+4}$ **$x = 0, x = 2$**

3. $\frac{5}{x+3} = \frac{6}{x+1}$ **$x = -13$**

4. $\frac{2x}{6} = \frac{x}{x+1}$ **$x = 0, x = 2$**

5. $\frac{8}{x^2-64} = \frac{1}{x-8}$ **$x = 0$**

6. $\frac{x+2}{x-2} = \frac{4}{x-4}$ **$x = 0, x = 6$**

LESSON 12-7

## Reteach

### Solving Rational Equations (continued)

If a rational equation contains sums or differences of rational expressions, multiply every term by the LCD to eliminate the fractions.

**Solve** $\frac{1}{x} + \frac{2}{x+3} = \frac{3}{2x}$.

$\frac{1}{x} + \frac{2}{x+3} = \frac{3}{2x}$ — *The LCD is $2x(x + 3)$.*

$2x(x+3)\left(\frac{1}{x}\right) + 2x(x+3)\left(\frac{2}{x+3}\right) = 2x(x+3)\left(\frac{3}{2x}\right)$ — *Multiply every term by the LCD.*

$2\cancel{x}^1(x+3)\left(\frac{1}{\cancel{x}^1}\right) + 2x\cancel{(x+3)}^1\left(\frac{2}{\cancel{x+3}^1}\right) = \cancel{2x}^1(x+3)\left(\frac{3}{\cancel{2x}^1}\right)$ — *Divide out common factors.*

$2(x+3) + 2x(2) = (x+3)(3)$ — All fractions have been eliminated.

$2x + 6 + 4x = 3x + 9$

$6x + 6 = 3x + 9$

$\underline{-3x \quad -3x}$

$3x + 6 = 9$

$\underline{-6 \quad -6}$

$\frac{3x}{3} = \frac{3}{3}$

$x = 1$

**Check:**

$\frac{1}{x} + \frac{2}{x+3} = \frac{3}{2x}$

$\frac{1}{1} + \frac{2}{1+3} = \frac{3}{2(1)}$

$1 + \frac{1}{2} = \frac{3}{2}$

$\frac{3}{2} = \frac{3}{2}$ ✓

**Solve. Check your answer.**

7. $\frac{3}{x} + \frac{2}{x+2} = \frac{4}{x}$
LCD: $x(x + 2)$
$x = 2$

8. $\frac{10}{x} - \frac{8}{x} = \frac{6}{x+1}$
LCD: $x(x + 1)$
$x = \frac{1}{2}$

9. $\frac{1}{4} + \frac{1}{4x} = \frac{5}{x^2}$
LCD: $4x^2$
$x = 4, x = -5$

10. $\frac{22}{(x+1)(x-1)} + \frac{2}{x+1} = \frac{8}{x-1}$
LCD: $(x + 1)(x - 1)$
$x = 2$




LESSON 12-7

## Challenge

### Exploring Reciprocals, Rational Equations, and Integer Solutions

In Lesson 12.7, you learned how to solve rational equations in one variable. Each equation below is a rational equation in two variables, $x$ and $y$.

$\frac{1}{x} + \frac{1}{y} = 1 \qquad \frac{1}{x} + \frac{1}{y} = 2 \qquad \frac{1}{x} + \frac{1}{y} = \frac{1}{2}$

In the exercises on this page, assume that $x$ and $y$ are positive integers.

**Find one solution to each equation of the form $\frac{1}{x} + \frac{1}{y} = \frac{1}{n}$, where $n$ is a positive integer.**

1. $\frac{1}{x} + \frac{1}{y} = 1$ — $x = 2, y = 2$
2. $\frac{1}{x} + \frac{1}{y} = \frac{1}{2}$ — Possible answer: $x = 4, y = 4$
3. $\frac{1}{x} + \frac{1}{y} = \frac{1}{4}$ — Possible answer: $x = 8, y = 8$
4. $\frac{1}{x} + \frac{1}{y} = \frac{1}{6}$ — Possible answer: $x = 12, y = 12$
5. $\frac{1}{x} + \frac{1}{y} = \frac{1}{8}$ — Possible answer: $x = 16, y = 16$
6. $\frac{1}{x} + \frac{1}{y} = \frac{1}{100}$ — Possible answer: $x = 200, y = 200$
7. Suppose that you are given $\frac{1}{x} + \frac{1}{y} = \frac{1}{n}$, where $n$ is an even positive integer. Based on your work in Exercises 1–6, write formulas in terms of $n$ for $x$ and for $y$. $x = 2n, y = 2n$
8. a. Let $\frac{1}{x} + \frac{1}{y} = 2$. Show that $x = 1$ and $y = 1$ are solutions to the equation by using substitution. $\frac{1}{1} + \frac{1}{1} = 2$

   b. If $x$ and $y$ are positive integers other than 1, then $0 < \frac{1}{x} < 1$ and $0 < \frac{1}{y} < 1$. Explain how this fact helps you show that there are no positive integer solutions to $\frac{1}{x} + \frac{1}{y} = 2$ other than $x = 1$ and $y = 1$.
   If $x$ and $y$ are integers with $0 < \frac{1}{x} < 1$ and $0 < \frac{1}{y} < 1$, then $0 < \frac{1}{x} + \frac{1}{y} < 2$. Therefore, $\frac{1}{x} + \frac{1}{y} = 2$ is only true when $x = 1$ and $y = 1$.

   c. How does the reasoning in Part b help you show that, when $n$ is an integer greater than 2 and $x$ and $y$ are positive integers, the equation $\frac{1}{x} + \frac{1}{y} = n$ has no integer solutions?
   Since $0 < \frac{1}{x} + \frac{1}{y} < 2$, there are no positive integers $x$ and $y$ for $n > 2$.

**Use trial and error to help you decide whether the given equation has any integer solutions. If so, state one solution.**

9. $\frac{1}{x} + \frac{1}{y} = \frac{1}{3}$ — Possible answer: $x = 6, y = 6$
10. $\frac{1}{x} + \frac{1}{y} = \frac{1}{5}$ — Possible answer: $x = 10, y = 10$




LESSON 12-7

## Problem Solving

### Solving Rational Equations

**Write the correct answer.**

1. Alex can remodel a bathroom in 2 days. Abraham can remodel the same bathroom in 4 days. How long will it take them to remodel the bathroom if they work together?
$1\frac{1}{3}$ days

2. When only the cold faucet is fully turned on, a bathtub fills in 8 minutes. When only the hot faucet is fully turned on, the same bathtub fills in 12 minutes. If both the cold and hot faucets are fully turned on, how long will it take to fill the bathtub?
$4\frac{4}{5}$ min

3. 100 mL of a solution contains 30 mL of acid; this is called a 30% acid solution because $\frac{\text{part}}{\text{whole}} = \frac{30}{100} = 0.30$. How many milliliters of acid $x$ would need to be added to the solution to turn it into a 50% acid solution? (*Hint:* Adding acid increases both the volume of acid and the volume of the complete solution.)
40 mL

4. Markus jogs 4 mi around a track at an average rate of $r$ mi/h. Then he cools down by walking 1 mi at a rate 3 mi/h slower. His whole workout lasts 1 h. At what rates did Markus jog and walk?
jog: 6 mi/h; walk: 3 mi/h

**Select the best answer.**

5. Ms. Spinoni can prepare a mass-mailing of 500 letters in 10 hours. Mr. Harris can prepare a mass-mailing of 500 letters in 15 hours. How long will it take them to prepare a mass-mailing of 1000 letters if they work together? (*Hint:* The complete job is twice as big.)

   A 6 h — C $12\frac{1}{2}$ h
   (B) 12 h — D 25 h

6. At a cafeteria, the automatic ice machine can completely fill itself in 20 minutes. During lunch time, customers can completely empty the ice machine in 30 minutes. At the start of lunch time, the ice machine is completely empty and it starts making ice at the same time that customers start taking ice. How long will it take for the machine to be completely full? (*Hint:* The customers *take away* ice.)

   F 12 min — H 30 min
   G 20 min — (J) 60 min

7. To attend a family reunion, Beth drives 100 mi from Fresno and Clara drives 220 mi from San Jose. Both women drive the same speed in miles per hour, but Clara's drive takes 2 h longer than Beth's. Find the length of Beth's drive.

   A 1 h — C $2\frac{1}{5}$ h
   (B) $1\frac{2}{3}$ h — D $3\frac{2}{3}$ h




LESSON 12-7

## Reading Strategies

### Use a Concept Map

**Rational Equation**

**Definition**
A *rational equation* is an equation that has one or more rational expressions. That is, the expressions have at least one variable in the denominator.

**Methods of Solving**
- Cross-products – when the equation forms a proportion
- LCD – multiply each term by the LCD to eliminate the fractions

**Examples**
$\frac{8}{x} = \frac{4}{x-1}$
$\frac{1}{m} + \frac{3}{m+2} = \frac{2}{m-4}$

**Extraneous Solutions**
Rational equations can have *extraneous solutions.* These are not solutions to the original equation. Solutions that are excluded values (values that make the denominator zero) are extraneous. Always check your answers.

**Answer the following.**

1. Which example above can be solved using the cross-products? $\frac{8}{x} = \frac{4}{x-1}$
2. Tell whether or not each of the following is a rational equation.
   a. $\frac{5}{6} = \frac{x-2}{3}$ no
   b. $\frac{4}{x} = \frac{2}{x^2-9}$ yes
   c. $\frac{x}{5} + \frac{1}{x^2} = \frac{5}{4}$ yes
3. What is the purpose of multiplying a rational equation by the LCD?
   It eliminates the fractions.
4. Liam solved the rational equation $\frac{x+4}{x-3} = \frac{3x}{x-3}$ and got solutions 2 and 3. Is either solution extraneous? Explain. 3 is an extraneous solution because it makes the denominators of the original equation 0, which is undefined.

**Solve each rational equation. Check your answer.**

5. $\frac{3}{m+2} = \frac{1}{m-4}$
7

6. $\frac{4}{c-4} = \frac{c}{4} + \frac{c}{c-4}$
$-4$